SCIENCE AND FREEDOM

LYMAN BRYSON

SCIENCE AND FREEDOM

Essay Index Reprint Series

BOOKS FOR LIBRARIES PRESS
FREEPORT, NEW YORK

Reprinted 1971 by arrangement

INTERNATIONAL STANDARD BOOK NUMBER:
0-8369-2385-5

LIBRARY OF CONGRESS CATALOG CARD NUMBER:
71-156620

PRINTED IN THE UNITED STATES OF AMERICA
BY
NEW WORLD BOOK MANUFACTURING CO., INC.
HALLANDALE, FLORIDA 33009

CONTENTS

INTRODUCTION

THE PURPOSE OF THIS ESSAY IS TO DISCUSS THE USE OF THE SCIENTIFIC method in understanding human behavior and the application of scientific knowledge to the management of human affairs. Such a discussion is not science, either pure or applied; it is a mixture of hypothesis with philosophy and argument. It is not addressed primarily to those who are building the new social sciences. If it helps them directly, so much the better, but my more immediate purpose is to help create the kind of moral and intellectual atmosphere in which they can do their work.

The thesis is that science can be used as one of the tools by which to establish the conditions in which human powers may realize human values. In order to come closer to the fact, I choose a value, freedom, postulating that freedom is a good and describing the conditions in which freedom can be made real.

What science can do, both science as description and science as knowledge in action, is indicated. In an historical sense, such an essay, whatever its merits, is a contribution to the development of a scientific humanism. There is now, in the superficial layers of the public mind, a momentary revulsion against science. That will pass. A cool appraisal of our resources, if it can be made free of institutional envy and party rage, will show that the most promising line of development now in social philosophy would have to be this use of exact methods to establish the conditions in which our powers can attain our values, a truly scientific and truly humane scientific humanism.

The words are not so fresh as they were a few years ago, but "scientific" and "humane" are still the best adjectives to describe our purpose. To struggle to better in every possible way the lives of all human beings is humanism. In so far as may be possible, this attitude will compel us in all our thinking to take persons as ends in themselves, not as means to the purposes of an abstraction such as a culture or a state. We can do this and still not forget that the personal and the social are two aspects of the same event, not in conflict but in collaboration. We shall try to be scientific by speaking never of the single or the peculiar event in nature but making our judgments on the discernible generalizations, the uniformities that can be seen in human behavior, trying to consider them without bias. Above all we shall seek to state our judgments in the public world where they are subject to any man's question.

The moral rules of our inquiry are the morals of science. Confusion may easily arise at this point. I do not mean that general moral issues are to be settled by appeal to scientific investigation. That is a matter worth discussing, but not pertinent here. Nor do I mean that science has any moral authority in the world at large; indeed it would appear that it has very little outside our Western groups and even at home it has not yet achieved enough success to be quite trusted. The point is rather that science, as one of the greatest of human institutions, and an honorable achievement of the West, has like other institutions a moral atmosphere in which its members live and for which its acolytes are trained. The first of these moral principles, already stated, is that all judgments are in the public world, in the open, where men's minds can try to agree, not in the arcanum of any single personal experience where any man can believe what he pleases. There are metaphysical pitfalls here of which we can be aware without letting them hold us. In the public world we assume that minds can communicate with each other and that they exchange symbolic references to the common world that makes their shared reality.

Another moral characteristic of scientific method, binding on our thinking here, is that no judgments will be given as absolutes. All judgments will be not only open to any man's question but will be offered always with the expectation that they may be changed by time and greater knowledge.

Also, all judgments will be secular, without prejudice or implication as to religious systems. Much further discussion will be needed before we can reconcile an operational scientific examination of ideals for the purpose of making a better world, not here and now but certainly here and as soon as possible, with the absolute requirements of religious faiths. At this stage of our argument, it will probably have to be taken on trust that the kind of thinking we speak for will not hinder religious development. A scientific humanism cannot be appealed to, however, in support of any program of inhumane behavior based on divine revelation. We can be concrete but we shall take an example that is safely remote from the experience and the religious thinking of a likely reader. Scientific humanism could not be appealed to in support of the ancient Hindu doctrine that a widow must be burned alive on her husband's funeral pyre, and no revealed truths as to the metempsychosis of an unburned widow would prevail against the humane conviction that this was not justice or mercy. But if knowledge, pursued in sweat and candor, knowledge of a world that must have been created or willed by the God of whatever religion one may hold, is not knowledge of the true nature of that God, the fault does not appear, I think, in the scientific method.

It is important to say also, since the word "scientific" has been given such curious political meanings, that the basic postulates in the attitude here taken are Socratic, not Marxian. By that, trusting that I am not doing great injustice to the adjectives, I mean Socratic in the sense of believing that man can examine the meanings of his own words and the import of his own beliefs and so can use intelligence to better his life on earth. He can change the social order by rational intention, instead of having to depend, in primitive Marxian terms, upon the slow turning of economic determinism. Without taking any position whatever on the question of economic factors in the determination of human behavior, we can still insist with Socrates that intelligence and purpose are real, not illusory, factors in human destiny.

Indeed, as an aside, I would say that Marxianism of the conventional sort is too easy a way out of the problem of evil. It releases too many of us from responsibility for the tragic state of man. There is too much escape in the idea that ultimate decisions rest on conflict between only partly independent and almost anonymous collectives, which are unconsciously controlled in their choices by decisions made in other areas. Spiritual choices, they say, come from production processes. The Socratic says that life has many incurable evils and man can only partly decide what shall befall him, but it is still true that much happiness and much misery are the natural results that follow on our acts. If we could behave differently we could have a somewhat different fate. The mistakes may be in our methods. Or they may be in our motives, that is in our ideals; and so ideals, no matter what their sanction, must be subject to examination by human reason.

We shall have much to say of the nature of freedom, but we take it for granted that freedom is a good. The postulate will be examined but not with any question as to its validity for our uses. Any arguments given in support of freedom as a value are not intended to be exhaustive or conclusive; we shall only discuss it as men discuss their absolutes, for explanation, not to justify. Freedom is a human good. It is created in the natural world by human effort, and it is self-perpetuating in the sense that it allows the creative powers of men to work at their highest energy. I believe that it requires a society of plural values where diversity is nurtured and where differences provide a matrix in which social change can be speedy and economical in both blood and labor but where change can be rationally controlled.

Our humanism is scientific because we believe in control of social change by intelligence and experience, directly and freely applied. In living by the scientific humanism that I am offering as a social philosophy for now, we shall use social engineering to solve the problem of setting up the conditions of freedom but not to determine what men shall do with freedom when they get it. Indeed, we are arguing here that the chief uses of freedom are defeated by those who set up its conditions and try also to determine its content. Here we shall speak for freedom, not in order that men may do what we predetermine for them, but in order that their freely working powers may create whatever they can plan and reach up to.

A color of derision and disdain has been cast over the words "social engineering" and "manipulation," sometimes by those who deliberately want to obfuscate the issue, sometimes by those who innocently think that all forms of eloquence are something else than manipulation. I shall need to say much more about the manipulations, the social engineering, of teachers and priests and poets and medicine men and prophets and creators of values since culture began, but not in derogation. They are the leaders who have invented values and taught us to believe in them. They have been the leaders in our spiritual progress. But my comment will be in strenuous opposition to any who object to fully rational attempts to better human behavior on the ground that clear-headed manipulations are somehow immoral, whereas other persuasions or other controls are righteous. It is my argument that the attempt to change the behavior of other men, or of men in general, is a normal human activity in which anyone who breathes and speaks takes part. It is to be judged not by its motives but by its means and ends. The means must be effective to good ends; in this case "good" is what leads to freedom.

Culture creates persons out of human potentials in ways to be described and the program of this philosophy is to set out deliberately to build such a culture as will create persons that love liberty and would be unhappy without it. Such a culture will create free persons and institutions by which they can live in freedom. Scientific humanism is a philosophy by whose principles we scientifically study and master human behavior for the fulfillment of human powers.

SCIENCE AND FREEDOM

WHAT IS FREEDOM?

THEY SAY THAT MODERN BOOKS END WITH DEFINITIONS INSTEAD OF beginning with them, and there could be some truth in that quip. One could take it to mean that when a writer has told thoroughly and adequately what he is talking about he has probably said all he knows on his subject. In spite of this, it is still useful to make clear, if we can, those terms that shine brightly in men's minds but are still tokens of misunderstanding. The question is not, simply, What is Freedom? To any one trying to be as exact as his material and his powers will allow, such a question is better put this way: What do I mean by the word Freedom? Or as a statement: The word Freedom is here used to describe certain conditions of human existence which are—

We try for univocal precision partly to avoid the mistake of discussing the content rather than the conditions of freedom. We need also to avoid the bewildering mazes of words into which men wander when they try to define freedom by Hegelian tricks, ending at last with some slavery by a pretty name, instead of what they said they were looking for. Beyond these reasons there is the specific one that has to do with our argument here. This essay has a practical purpose.

A trial definition, foreshadowing part of the discussion, is to say that the technical term Freedom will be used to name the social conditions in which there are enough normal choices of behavior patterns open to every person to allow for experiment, and change, and diversity, both in the successive experiences of individual persons and also among different persons in the group.

It would be possible at this point to take up the ancient problem of the reality of the freely choosing will, but with the metaphysical aspects of that question we have nothing to do. In my opinion, the philosophic statement made by Charles W. Morris is enough for our purposes. Morris says:

> Thus the conduct of the organism is in part a function of things indicated by signs. A new conception of freedom begins to emerge: it might be said that the animal is "free" from his immediate environment in the sense (and to the degree) that its conduct is not solely a function of its immediate environment. "Free behavior" begins to mean "behavior directed by the operation of signs."
>
> There is no absolute line to be drawn between the behavior of an animal that perceives and the behavior of the conceptually thinking human being; it is a terminological matter whether all sign-controlled behavior is to be designated as free behavior or whether this term is to be applied only to the most complicated phases of such behavior. Nevertheless, the differences, while differences in a continuum, are important, and there are grounds for preferring to restrict "free behavior," in the third sense of the term, to such behavior as involves the symbolical indication of alternative courses of action and the selection of one such course of action in and through the process of symbolical indication. . . .
>
> For in speech with others and himself man is easily able to keep the sign operative independent of the immediate environment; to bring before himself for consideration (by the interrelationship of signs) the consequences of a mode of action; and to try out experimentally alternative modes of action by stimulating himself through various sign combinations. Thus as a result of sign functioning at the linguistic level the human being reacts to a world which extends indefinitely beyond his immediate environment and his perceived environment, and is able to lay out courses of action after consideration of the consequences of various possible alternatives.[1]

In any case, the scientist and the statesman, the teacher, the social engineer, even the poet and the priest, are compelled for secular purposes like ours to take men on their overt behavior. Men "act as if" they had free choice. The philosophic question, whatever might be discovered as its answer, could not change

[1] *Freedom: Its Meaning*, ed. Ruth Nanda Anshen (New York, Harcourt, Brace, 1940), pp. 583-584. Reprinted by permission of the publishers.

either practical or theoretical considerations in this argument. We will speak of men as free agents and of the conditions that enlarge the uses of their purposeful strength.

In a group of free persons, there must be a range of choices and that range must be known. Knowledge concerning what can normally be done must be easily accessible to all persons. I have elsewhere written of the bearing that knowledge and freedom have on each other.[2] It is clear that men are not free to do things they have never heard of, and that knowledge enlarges their freedom not alone by telling them of choices but also by making them aware of the likely consequences of each choice.

Also, the choices opening out ahead of a free person in a free society at any moment of suspended judgment when he is, as Morris says, rehearsing by symbols his contemplated actions, must all be within the patterns of the "normal" for that time and place. Otherwise, his freedom is an illusion. Many kinds of action may be within a man's physical and mental powers, but he is not really free to do them if in doing them he would be called criminal, or an outlaw, or eccentric, or ludicrous.

A free society is one in which there is a rich variety of normalities, and this is a crucial point in thinking clearly about freedom. Simple societies are seldom free in this sense although complexity is not in itself to be confused with the conditions we are going to describe. A society or a social group that treats with the same, or nearly the same, friendliness a wide range of behavior patterns is providing freedom in some real degree for its members.

It should be noted that we do not speak of original nor of eccentric behavior. Whether these are desirable or not, and under what conditions, are other questions. Freedom, as provision for free "normalities," lies in the establishment of patterns of wide range and flexibility for the person who does not have the power

[2] Lyman Bryson, *The New Prometheus* (New York, Macmillan, 1941), p. 107.

or inclination to be original but who does desire to contemplate several courses of action and follow his mind.

And also it should be noted that we are speaking only of choices in matters of some importance. As we shall see, the person is here considered as constituted of habits in all ordinary behavior; that is what makes one "ordinary," or predictable. This truth is often forgotten in training a child for liberty; we ask him to choose in a vacuum, unprovided with knowledge of things to do. The personality is a mosaic of habits, but creative factors are nevertheless at work in it and besides, the crises of life, coming on with change and age, make new habits useful. In a free society, men do not feel under any compulsion to be all alike nor even to remain as they are. On the other hand, they do not carry a heavy burden of constantly having to invent, each for himself, wholly new ways of doing necessary things. It takes strong persons to be free, and free societies make for strength. But that strength is not to be spent on minor matters.

The definition speaks of "all persons" as having access to the range of choice that makes liberty actual. We ought to consider the grave question, Can a master be free of the bondage in which he keeps his slave? At this point, we are making a more simple statement. One measure of the freedom in any group is, like the measure of any other value, the number of members who enjoy it.

In any social system, time and circumstances will make all choices difficult and all consequences uncertain. The arrangements of a culture cannot indefinitely enlarge either the freedom or the powers of its members. Besides, there are always serious barriers between man and the possibility of making further new choices after a course is embarked on. For one thing, no man can ever go back to any earlier parting of the ways in his own life and decide to try again and do it differently the next time. He is not, and never can be again, the man he was when he made his first choice. A free society is one that does not increase these difficulties unduly. It will not specialize men into a slavery that is the forced continuing of what was once free action.

From the anthropologists [8] we know that most men can be normal in some one of the normal patterns accessible to them in the time and place in which they happen to be born. There are always deviates. There are always temperaments that do not fit a small range of patterns. One of the reasons for having a larger range is to reduce the number of persons who find themselves marginal to all normalities and thus must be eccentric. A free society would not be aware that there were many deviates, because in such a society only a great degree of eccentricity in an individual temperament could compel a person to act outside the range of accepted patterns. This is not the only reason for having the range of choice, however. The ideal is not a society for eccentrics, nor a society so diverse that it has no form and no possible pious continuities. A further reason for the range of choice, both horizontally among the members of groups and vertically in the successive changes in the life of an individual, is that such plural possibility makes the social change that is inevitable more intelligent and more peaceful.

II

The conditions of freedom are the conditions in which the creative powers of men, and all the other creative factors of nature that man has access to, can work best. If this is a desirable state of society, a social good, the rational or deliberate attainment of it is of preeminent concern. This essay undertakes to suggest some of the means of attainment. I agree with those, now growing slowly in number and authority, who believe that the behavior of men, in all cultures and at all times, has uniformities that can be scientifically described. They can be manipulated and controlled as we control other natural phenomena by similar investigations and similar engineering skills. If we can control them with a real desire for freedom and creative change, we can build social structures that will make creative change not violent and costly, as it has almost always been, but peaceful.

[8] Ruth Benedict, *Patterns of Culture* (Boston, Houghton Mifflin, 1934).

The social structures of a free society will have to be complex, however, and will allow for pluralism. I believe that men can achieve freedom only if they can learn to compose their differences of temperament and belief and cease trying to destroy their differences by destroying each other. Many ways have been tried to get agreement on the basis of absolute doctrine and the history of all these trials appears to show that if men in power believe that truth is absolute in its accessible forms they can either have truths so general that differences do not matter and the intellectual life is inert, or else they can, by suppressing all dissent, get the look of agreement behind which differences, if they arise, do not get expressed. But by using intelligence and good will, men can organize societies of not one but many different kinds, in which there is agreement as to the needs of unified action whenever actually useful. With that unity in action, there can be an affectionate and encouraging nurture of real differences for the sake of their creative power.

The doctrine of free progress, however, if by that we mean real change, is not altogether a kindly one. It must be harsh in judgment on many things as they are now. Even more, and this is not often understood, it must be harsh toward the wishes and plans of such men as we ourselves are now. A real doctrine of progress will be implemented in action when we work rationally for a future that we know will not be in any large measure what we now expect. As Margaret Mead has said, in discussing the use of science for the achievement of social progress, "In order to implement a spiritual future which transcends our present cultural values, we need humility to realize that we would no longer be at home in such a world; that we who have dreamed it could not live it."[4] The social scientist, if he is honest, must be a kind of spiritual Hippalus; he must be like that mariner who was first, the legends say, to put his ship out beyond sight

[4] Margaret Mead, "The Comparative Study of Culture and the Purposive Cultivation of Democratic Values," in *Science, Philosophy and Religion* (Conference on Science, Philosophy and Religion, 2d Symposium, New York, 1941), p. 67.

of shore, trusting not to his immediate vision but to his science and the stars.

We are ourselves the present generation and we cannot dissociate ourselves from those stages in human behavior that are to be transcended except in our willingness to be parents as well as dreamers of the future, to live in that rhythm of death and sacrifice by which new life comes on. If we accept the declaration of Godwin (*Political Justice*), "Nothing can be more unreasonable than to argue from men as we find them to men as they may hereafter be made," we must count ourselves, honestly, among the men now found. It is necessary to insist on this because most men would appear to believe in progress only in so far as it means a future dominated by more men of their own kind. They will not face the logical and moral necessity that posterity shall not be made up of our friends but must be an unpleasant multitude of men in many ways unlike and hostile to ourselves.

Does this then mean that we must work for change blindly and without real direction or, worse, that we lose in the struggle the motives and values that impel us to make the effort? That is not the fact, nor my intended meaning. We can work for an ideal of freedom which is by our own definition a larger and more creative force than any man or any man's ideal can be in itself. And since all behavior is an act of faith, we have faith in the creative power of men in the freedom we can help make possible. We can do it by the scientific manipulation of all the factors inside man and in his environment that give play to the intelligence, and so our faith is actually in intelligence itself, not only in its past achievements, nor even only in those that give us our present powers, but in creative change.

This is to act in regard to the most important aspect of our lives, our own behavior, as we have successfully acted in regard to lesser concerns such as the control of material things and the other animals. The problem is not to find a new scientific mode but to find the entities in human behavior that can be scientifically used.

Before that is done, it may be useful to explain further how my purpose differs more than may appear, at first, from the purposes of other writers on freedom. It hangs somewhat on one's idea of social control. Most writers on freedom, doubtless without knowing it, follow the Platonic [6] rather than the Anglo-Saxon idea of law.

They do not devote themselves to devising the conditions of freedom because they are too much concerned to declare what men are going to do when freedom comes. They have too much to say on the content of that freedom to be concerned with its conditions or techniques. They want men to be free to do the things they would prescribe. Thus our Puritan forebears wanted freedom to worship as they pleased, but had no idea of founding a commonwealth where any who disagreed with them could enjoy any freedom whatsoever. The Puritans believed in their ideals, no doubt, and one of their freest acts was to hang a disruptive Quaker if they could get their hands on him. Too many writers on political freedom speak with more spiritual affinity with this idea than they realize. They love no diversities for their own sake; liberty is not desirable to them for the very reason that it will permit men to do in their creative selves such deeds as cannot possibly be anticipated now. But that undetermined liberty is precisely what I would provide for in social conditions and controls.

A concern with the conditions and techniques of freedom and a determination that the content of that freedom, the behavior of men in it, shall be determined by their own progressive creativeness, does not mean that liberty has no limits. It does mean that "planned" societies, including this one we are planning now, are all brought into question, and all Utopias are suspect. We are

[6] Huntington Cairns, "Plato's Theory of Law," *Harvard Law Review*, LVI, No. 3 (Nov., 1942), 360; O. W. Holmes, Jr., *The Common Law* (Boston, Little, Brown, 1881), p. 41.

in this point disputing not only modern reformers but Plato also, on high grounds. In this, as in so many other phases of his endlessly suggestive thought, Plato could not know how later experience would disentangle phases of being which to him were the same, and would analyze doctrinal fields wherein he thought he was dealing with one subject rather than with many. Legislation and education were so close together in his mind that the characters of men were to be determined in the greatest possible degree by the state. It is for that, perhaps, that he seems to so many modern thinkers[6] a "Fascist," but he was, in fact, doing only what any political thinker or educator or "planner" does who tries not only to give men powers and choices but also to determine their use of them.

The limit on the content of freedom that I shall make here is only that it cannot be used to limit itself. The society that I am proposing would be deliberately constructed to make men happy if they loved freedom, and afflict discomfort or misery on those who so far escaped the natural effects of freedom in education as not to want freedom in life. And this is to apply in political ethics an idea from the common law of British, American, and Germanic traditions, out of Rome in some measure, perhaps, but not out of Plato and Aristotle. Justice Holmes expressed it in a letter to Pollock (September 1, 1910). He dismissed a presumptuous writer because he failed "to understand the purpose of the law to fix a line of minimum social conduct required of men at their peril."[7] The Greeks, on the contrary, wanted to establish the ideals of men in the sentences of magistrates. What we are speaking of is the use of social control, not only explicit legislation by political bodies but all forms of social influence and direction, so as to make it possible—even necessary—for many men to be always seeking higher ideals than their group has as

[6] R. H. S. Crossman, *Plato Today* (New York, Oxford University Press, 1939).

[7] *Holmes-Pollock Letters*, ed. Mark DeWolfe Howe, Sept. 1, 1919 (Cambridge, Harvard University Press, 1941), I, 168.

yet come up to, and for all men to live by ideals that call out all their powers.

In one sense, of course, we are only saying that Plato's ideas of law, and of freedom too, although wise and immortal are too primitive for us now, after 2,000 years of thought and experience. His thinking took for granted that there is one way of the good, the true, and the beautiful, and that society in shaping men also determines them. So also thinks the savage, although his idea is brutish while Plato's is full of light and greatness. We have gone beyond—with the help of both Plato and the savage.

Even as early as this, in any discussion of more exact methods to deal with human affairs, one can feel the hot breath of an old dragon. In spite of all our talk of "freedom" as the value worked for, is what we are about to propose only a way of putting man into the strait-jacket of "scientific determinism?" There is too much serious work to be done for us to stop and argue out the question as to whether or not there is such a thing as "scientific determinism," outside the imaginings of unscientific minds. Science can scarcely be deterministic in the philosophic sense since its whole purpose is to discover such knowledge of observable events as will give men the power to predict them and, by predicting, master them.

The ghost of meaningless controversy, however, is not so easily laid. It will probably take the whole argument of modern scientists willing to enter the lists to show that all knowledge, scientific knowledge like any other kind, makes for freedom, and that man cannot ever lose his own self by knowing more about the world in which he acts. But there is reason to hope that we may come at last to a kind of Socratic morality, after all our two millenniums of backsliding, and believe that knowledge and freedom are two adjectival aspects of the same good life, the same state of being. In such a development, of course, knowledge will not be contemplation merely but practical knowledge that gives men help in action.

We are here describing the conditions of freedom as present cultural accumulations make those conditions known. But our escape from the philosopher's problem of free will cannot be quite so simple as what has contented some men. We cannot, like Thomas Hobbes, merely say that "A Free-Man, is he, that in those things, which by his strength and wit he is able to do, is not hindred to doe what he has a will to." [8] This will not do for us because we know that the "will," the final impulse to act, is conditioned by such knowledge of possible choices as may be in the mind that chooses. It is true, I think, as Hobbes thought and as anyone is likely to think who is not dealing in meaningless words, that the man and his decision are the same thing. There is a kind of interior coercion. But it does not lie in that impression of being forced that comes from the conflict of impulses that are really the factors of one's self. It is the coercion of ignorance. No man, as we have said, can choose to do what he never heard of doing or never thought of doing. In this sense, the ultimate measure of freedom is knowledge and we learn in order to be free.

It is an observable fact, as evident to our sensory examination as is any other physical fact, that individuals make changes in the world. If this way of putting it is philosophically not acceptable, it really means the same thing to say that changes take place within personalities and these changes are followed by changes in physical events. And this is creative action. By conditions of freedom we shall mean, all the way through this essay, those conditions that make the greatest possible opportunity for creativeness to work. Our whole purpose in seeking the conditions of freedom is to give man's creative powers their chance.

III

The modern forms of freedom are usually called democracy; we can get clearer ideas of what we mean by freedom if we

[8] Thomas Hobbes, *Leviathan* (Everyman), p. 110.

discuss what we mean by this other word. To the modern Western man freedom presents itself as something to which the title of democracy belongs, and he could not easily be convinced that any other guise of freedom could be honest. But the word "democracy" has been obscured by overlapping interpretations. We have no use for any such term here unless we can give it a single meaning and that may be done by dividing the idea into kinds.

First, we should note that what are taken for disputes as to degrees of a badly defined condition—that is, "democracy"—are, in fact, disputes as to the kinds, and that there is no certain proof in political history that the kinds are cumulative degrees of one condition. It is quite conceivable that some kinds of democracy can be achieved in high degree in societies where other kinds are missing, as in those periods in the United States, for example, when we have had the political kind without the economic kind, or as for long periods in China, where there has been the social kind without the political.

At the same time, we may say that the kinds of democracy, although they are not degrees of cumulative achievement in freedom, are at least so related to each other that the achievement of any one will make getting the others more likely and easier. This is in part because men take for granted the liberties they have and use them as familiar stepping stones to the others that can be imagined from each vantage point.

In this country, we have gone furthest in attaining the political kind of democracy; in fact, we have it in a degree not surpassed elsewhere although some of the newer countries, Australia or New Zealand, may have better machinery for registering political decisions of freely acting majorities of their citizens.

The central purpose of political democracy is not the most efficient registering of majority wishes, however; nor is its central purpose the most efficient management of public business. Here is a crucial point that can be grossly misunderstood and can cost peoples their liberties and their government's efficiency as well.

The essential point in democratic political philosophy is that government is not an end in itself, nor the state's business the dominant business of the state. A democratic government is one that has for its purpose the creation of such conditions as will best keep and develop the intrinsic powers of men. Our forebears who built our nation, somewhat more consciously and philosophically than men have usually had a chance to do, understood this matter very well. They knew that in a democratic government the citizen is an end in himself and that all governmental operations are for his advantage. That advantage is in those institutional surroundings that will give the best of him the greatest scope.

In this sense, government is primarily, and not merely incidentally, an educational agent. The government educates by its own self-denial; the Jeffersonian idea, that the best government is that which governs least, is rooted here. The best government governs as little as possible, because the best government puts the greatest possible weight of responsibility upon the individual citizen. We are educated by what exercises our powers; we grow in the environment that demands of us as much as we can give. If we can keep the discussion of this issue out of the intentional obscurities of practical politics, in which words and the ideas they purport to represent are used as weapons and not as tools of thought, we can realize that the line between public and private responsibilities is a very delicate one, hard to draw or to keep clear.

Even Adam Smith, on whose shoulders most of the blame for laissez faire has been piled, said that governments should intervene in economic situations to prevent injustice, and there seems to be nothing in the *Wealth of Nations* that justifies a modern corporation, legally masked as a "person," to claim the liberty to exploit. There are situations, conceivable at least, in which the government must intervene between a private group and the public, that is, between one small set of its citizens powerfully organized, and other less well-weaponed citizens, in order to prevent the same suppression of individual choice for which we must often fight against government itself. And equally, there are situations conceivable in which intervention by government, as in requiring children to attend school, is necessary to keep up that knowledge which is itself a vector of freedom.

No such cases should obscure the fact that government, being by definition the exercise of power, tends to enlarge itself, growing by its own success, and that too much government applied in the wrong places is evil. Bertrand Russell discusses with insight and courage, in his *Freedom versus Organization*, the dilemma developed in the nineteenth century when we built structures of industrial and engineering efficiency, institutions that put our science to work, while at the same time we were trying, with somewhat less success, to keep and develop the older ideas of freedom in which the person, not the material product, was the end. Democracy is that form of organization which has the person as end and serves the purposes of the individual by the educationally effective demands which the work of the organization puts upon the person's strength.

It is as true now as it was when Plato said it that a democracy is a form of government in danger of corruption. And, in spite of the fact that much of the Fascists' efficiency has turned out to be a hollow pretense, it is true that democracies are not always more efficient than autocracies by mere mechanical standards. That is, they do not always, nor even often, get their business done with notable dispatch or cleanness, if by their business we mean the devising and enforcement of the laws. It is necessary and not impious to criticize our democratic forms of government for their corruption and their inefficiency, but proving "sins" of that order does not make a valid case for revolution. A democracy has lost its claim to the support of patriots only when it no longer conducts public business with the help of a maximum number of its citizens and when it no longer makes the enlightened participation of those members its chief concern.

Herein, evidently, is the real evil in the continued centering of power. In the United States, not only in Franklin D. Roosevelt's administration and the New Deal, but before his time, in the administrations of Herbert Hoover and others, the mere size and complexity of modern government have seemed to require a drift to the center. For this trend, in order to make it palatable and even ideally exciting to as many voters as possible, the term "planning" was adopted. The ineffectiveness of much previous governmental management and the evident fact that, in a country as large and variegated and complex as ours, no one was thinking much of the country as a whole, gave reasonableness to the plea for "planning." The popularity of the term, sloganwise, was a great aid to the useful experiments in public works like the Tennessee Valley Authority which might have been blocked or slandered into desuetude if the people had not been roused to admire "planning." In these achievements, the cult of planning had great justification, in spite of a good deal of nonsense written by busy central government agents to whom planning obviously has meant not the organizing of national thinking but the imposing of standard means for standard ends. Enterprises like the TVA are attempts at organizing government interventions in such a way as to escape the bureaucratic evil; in fact, they seem to have aided the devolution of public burdens by increasing the share in the directing of his own destiny that each citizen was asked to carry.

But centralization of power, as such, is evil in spite of the fact (in so far as it is a fact) that by such gathering in of power we can get public business better done. When a decision he could make for himself is taken out of the hands of a citizen, his government has failed in democratic operation. It does not justify to say that the central authority makes a decision that does the citizen material good. The benefit that a democratic government is seeking for him is produced by and in the decision-making process. Upon this as upon all governmental principles there are evident limits. Decisions must often be made on behalf of others in all phases of communal existence. But democracies, as long as their leaders know what they are doing, sacrifice everything possible to the principle of self-government for its own educational effect.

For a long time, we have believed that the enlargement of the units of government administration that have seemed necessary to control size and complexity could be fully compensated for by increasing the political knowledge of private citizens. The techniques of adult education have been studied in the hope of finding a way of making political affairs again more interesting to voters. It is possible to say "again" because we have so much evidence, from writers about American life in the formative years of the nation, to the effect that politics was an absorbing and incessant topic of American talk. They said we talked of almost nothing but our public affairs. That high excitement, which may have merited satiric descriptions by foreign visitors, who wrote often as if they thought they were watching babies set up self-government in a nursery, was crude but real. One cannot be sure that it was not its reality as much as its crudeness that made some of our visitors smile. There was not much faith in democracy anywhere but in America until a time when the United States had become not so much less crude as more powerful. From all this we have grown into indifference, and tend now to satirize our own politics without asking the help of foreigners.

To remedy this lack of political conscience and revive political concern, the techniques of adult education have not, as yet at any rate, been enough. Public forums have been established and have been as widespread and as well attended as the lyceums of the 1830s. Newspapers and radio and documentary moving pictures, many devices in the hands of honest and devoted teachers, have achieved much but not enough. It is now too soon to say that later results may not be more satisfying. We are here urging that old techniques are to be improved and new ones invented. We must act on the principle that knowledge is the actual source of freedom; but we need not on this account fail to see the other factors in the political situation, which prevent us from keeping the public enough informed to enable the average voter to cope with bewildering public business. Our present political behavior is not so much a valiant although hopeless struggle by patriots to meet their political obligations as it is a sourly humorous indifference; in that mood education is mocked and the purpose of democratic government defeated utterly.

It is said, of course, that the citizen is bewildered because the problems are too much for any layman's information or capacity. That may be true, but it is another example of slipping away from the true purpose of democratic government. The question is, What would be the effect on the citizen of having him *try* to accomplish these almost impossible public tasks? If it is good in the long run, much loss in the effectiveness of getting the business done is a small price for the nation to pay.

The trend toward central power detaches the citizen from his education, cuts him off from the experience which democracy was created to provide. It is not true, I think, that the size and complexity of state problems have been the cause both of the trend toward central power and the defection of citizens from self-government. The process seems to have moved through intermediate steps. Concentration of power, when it relieves the local citizen of responsibility and a share in public business, thereby cuts off his interest together with his responsibility and his participation. No amount of mass education can do much to keep up political interest in citizens to whom government has become remote, paternal and out of reach. What is needed to revive that concern is to do all we can to spread information and understanding and also to check if possible the centripetal flow of power. It is difficult to think of any government which has ever peacefully given up central power, when once acquired. But it is politically immoral to follow a trend merely because it is evident as a trend; just as it is immoral to approve or abet any other fact for no other reason than that it is a fact. Political morality, like all morality, lies in positive choice.

If the trend toward central power is irresistible, then sensible political action would be to save democracy if possible in spite of it; in other words, to use power to spread knowledge and to maintain the structure and function of local, personal action. Many schemes of the New Deal between 1932 and 1944 were by this standard vicious, many of them were helpful. It may have been because we are so little interested in politics that we failed to welcome one kind of "planning" and to repel the other kind.

Beyond these considerations, the larger fact remains that in the United States there has been and still is a deep loyalty to the institutions of political democracy, and a reality in action. Other kinds of democracy can be usefully examined in the light of the same principle. Do we have social, or economic, or cultural democracy, if these kinds are to be defined as various ways in which institutions can be established to make demands upon persons and so educate them and make them great?

Much of the talk about these kinds or, as they are usually taken to be, the degrees of democracy, is kept from cutting deep into the truth of the subject because of careless definitions. When political power is at stake and passions run high, there are no definitions at all. It is forgotten that the purpose of democracy in government is the development of free human personalities by means of participation in public business. The basic implication is overlooked that any other kind of democracy will find its reason in persons and not in things or material results. The person as a possibility of excellence is still the end and cannot be a means to other conceivable goods. On this account, social democracy is not for the sake of graceful or beautiful manners but for the human virtues that will be developed in all the members of a society in which there is no caste. Industrial or economic democracy is a system that uses the processes of material production and distribution to enhance the dignity and happiness of workers and this is not the same thing as to raise the standard of living. The cultural democracy that will come in good time will be the social use of all resources to widen the share that each member has in the enjoyment of the institutions that create value, and even to increase the share he has in the creation of new values and new institutions.

The lack of social democracy, the existence of a hierarchy of social subordination, is a handicap to the establishment of freedom and a drag on the work of creative minds. But other kinds of democracy have been magnificently achieved in some countries without much of this kind, as in Great Britain, for example. It has been possible in Britain to make men free in many aspects of their lives, to remake many sets of institutions in such a way as to direct them to citizens as ends, without tearing down feudal social structures and symbolic rites that came from the older day.

On the other hand, in a country like the United States where there never were any well-established feudal habits, social democracy goes naturally along with the other kinds and threats to it come from the failures and lacks of the other kinds. This is easily seen in the social mobility of men who can move up or down on the money income scale.

Economic democracy is more talked about than social and is certainly more important. It is also more generally misconceived. As the term is used, it appears to mean at least three different conditions: one, equality of opportunity; another, equality of income; and by my own analysis here, economics for the sake of human development.

There ought to be something in the Marxian analysis that would help us, but I have not been able to find it. Perhaps it would be more precise to say that what I am saying is implied in Marx and his earlier commentators, but has been obscured in the practice of Russian communism. The evidence is convincing that the individual members of the Russian society have been sacrificed ruthlessly and often to material production and to reasons of state. If we ever see in Russia the withering away of the state and all state coercions, when the creative powers of which there have been magnificent flashes in twenty-eight years of swift experiment are loosed, Russia will then have some kind of economic democracy.

It is important to point out here, however, that what I am looking for in Russia is not political freedom. The claim of those who fight Russia's international battles all over the world is that the present communist state is economic democracy and that political freedom will come later. The point is subtle and difficult. The objection to state coercion at issue here is that it is being used by Stalin and his hierarchy of political dictators to expand and maintain an economic system which is no more democratic than many phases of capitalism and less so than some. State coercion is being used in Russia to set up an industrial order which still has material achievement for its goal. To meet this objection it will have to be proved that the present Stalinist state gives first, not incidental, consideration to the effect of work experience on the worker.

IV

The other meanings of the term are not useful here. Equality of economic opportunity is a desirable social condition and it is true that we have lost in this respect some of the advantages of our earlier times. The maturing of the economic structure and the settlement of waste lands are sufficient causes to explain the loss, without imputing evil motives to politicians or businessmen. It is worth our striving to get back again as much as our present complex economic structure will allow.

Equality of income, the other so-called economic democracy, is doubtful, even as an ideal. The income scale as a means of social reward and punishment has easily discerned values. No modern nation ever took the ideal of equal money incomes very seriously for long, not even Russia.

The objection, by our principles, to both equality of income and any drastically contrived equality of opportunity would be that they make the material results rather than the human results of the economic process the primary aim. Of course, reasonable men are often inspired to achievement by the chance to get material advantages for their families, such as comfort and education and protection. We might even go so far as to say that the highest values in modern civilization can be realized only in a rich society; medical care, for example, and this has to be paid for, either by ambitious and successful members or by society as a group. But neither equality of income nor equality of opportunity is the essential point in a mature conception of economic democracy. We shall enjoy that condition only when we have redesigned our economic processes, our production and distribution, in order to make the worker's life more demanding, more educative, more enlightened, and when we sacrifice, if necessary, the quality and quantity of material goods to that end.

Whether or not such a change would inevitably demand material sacrifice is a very difficult question and we do not as yet have the experimental data to answer it. The philosophy of trade unionism at present, the only leadership that might naturally be concerned with this question, has not faced it, in so far as public statements indicate intentions. Labor leaders seem to believe that work is essentially evil and that all a worker could want would be shorter hours and higher pay. In spite of testimony from sympathetic and qualified observers that other factors often outweigh these obviously desirable things, and that the self-respect of the worker is often more important to him than his pay envelop, labor leaders seem afraid to shift any of the emphasis in their demands.

Labor leaders do ask that workers be given a share in management, but the reasons for this demand are, according to our principles, the wrong ones. They ask that workers have a share in management in order that the same material rewards—high pay and short hours—may be more secure. They do not, except possibly in rare and little known instances, make the basic claim that the industrial process is for the sake of the worker, not for the sake of either profits or material products. The worker's ultimate advantage lies not in his wages but in himself, and if he ought to have a share in management it must be because management is a phase of the process that demands more effort, more thought and energy and devotion than does labor; and hence, by demanding more, does more to develop the powers of those who work at it.

Much in the past propaganda for their own cause now stands in the way of a forward step for labor's leaders. They have not been willing to admit that management required so much greater effort that it not only deserved greater rewards materially but by its processes made better men. They recognized the principle by applying it to their own leadership, of course, and in the management of union affairs. They have given generous material wages to their own leaders and have sought participation in union management for the sake of the enlightenment to be got from the experience. But they have not demanded that the same principle be applied in industry itself. Management of unions is not management of the economic process, it is only a corrective, at present necessary, for imperfections in the process of production.

Economic democracy in the real sense would be nearly as revolutionary for trade unions as it would be for old-fashioned management and for owners of capital. The Whitley Council movement in Britain is a step in the indicated direction. Any committees of workers that understand the principle can work toward it. But as long as both management and labor serve the interests of the stockholders instead of their own, as long as they work only for more and better products, no matter how the products are made or how they are distributed they are not moving toward the kind of economic democracy that can have ultimate social value.

There may be much more of this idea in both the New Deal labor policy and in Stalinist Communism than I have been able to discover. For the purposes of this argument that point is unimportant except for any regret that I may have over doing political movements an injustice. The important point is to recognize that democracy in a man's working life, like democracy in politics or in any other phase of human action, is not a matter ultimately of power or rewards or the uses of power, but of growth in human personality to which all these other goods are only instruments.

Wages can contribute to the good life but they do not determine it. The right of the capitalist to his return is measured by the same social criterion, namely, What virtues in him are encouraged by the interest and the profit that he takes out of his investments? If his gains are no encouragement to his own human improvement, if they do not make him a better man by some reasonable social standard, he may have legal rights to them but those rights are not democratic. It is at least possible that there might come a time when the rights of a person, as an investor, would be diminished on behalf of other rights in which he would share as a citizen and which were essential to some achievement of real democracy. In other words, the laborer is worthy of his hire provided that hire makes him worthier. That is the principle of economic democracy and it applies to every person who presents a claim to a share in the social wealth, whether laborer, union leader, manager, owner, salesman, farmer, miner, professional worker, or servant.

It would be difficult to say that we are moving toward such a condition. A profound transformation in our ideals and purposes would have to come first. I do not believe that such a change can come until we have had a long period of free development on a lower level, until inventors of social norms have had a chance in freedom to tear our attention away from material goods as the final standard of happiness and get us back to ideals more ambitious and more humane. Since there is no reason for believing that the new ideals would cost us great material losses—and it is not even certain that economic processes democratically managed would be less efficient than those managed by an elite of gifted administrators—the risk is not too great. But a generation of men that cared more for themselves as men than they did for their possessions would take all the risks involved and would achieve their economic democracy by using economic machinery, not by being servants of it.

When we have attained democracy in social relations, in government and in work, we can begin to plan for other kinds. There is no evident reason for supposing that growth will take the direction I have indicated; it can only be said that I am urging this kind of progresss as desirable in accord with the principles of valuation here set up. It should be noted that even here we are discussing the conditions of freedom and not its program. The argument is that freedom, in this particular case the various forms of freedom called democratic, will be attained if man and not materials or efficient processes is made the end, and if all strength is applied to man's development. The direction of that development is left to man's creative powers. The only absolute is freedom itself; this was our postulate.

There is a paradox involved, however, of which we must take notice. Social contact, among all kinds of men, appears to be a social good because it is enlightening, enlivening, and pleasant. Social democracy will result in more of this social good as well as more variety for all who have a share in it. Political democracy will probably increase the scope of politics in the lives of all who take part, because public business will be carried on in talk, both wide and deep, wherein many citizens may speak. By using public action in such a way as to make it a good in itself, we shall probably increase political activity as well as political interest. But here is the paradox: there seems to be no reason why we should expect the same tendency to show itself in economic action. There is no reason for believing that economic democracy will inevitably create conditions in which most men will find that they are giving more time to the material needs and ambitions of life. On the contrary, we may happily discover that we can greatly lessen our concern for material things when we have learned to make material processes into tools of greatness rather than ends in themselves.

One further kind of democracy that can be foreseen as possible and hoped for is what may be called a democracy of culture. Here, also, it would be easy to disregard the principle and suppose that we had gained a democracy in civilization when we had provided all members of the society with reasonable means of enjoying all that his society could offer. This would be a social gain of the greatest importance; we can scarcely call our civilization great in democratic ideals as long as circumstance shuts men away from knowledge or from the exercise of those higher faculties by which we learn to use not only knowledge but all the arts. This free access to the stores of civilization is necessary. But it is not enough. We shall have a democracy of culture when the largest possible number of members of the social group, limited only by innate ability, take part in the actual criticism and creation of the culture by which they live.

The inventive mind is rare; or, at least, it has been rare in the circumstances of the past. We have no reason for supposing that we can breed greater men and women by any known method of cultural eugenics. But it is also true that we have no way of knowing how many great men and women have been bred in the past and thrown away. Indeed, we know that the waste must have been prodigious. It is true that waste is the way of nature. One of the reasons for feeling that the conflict between human purpose and the physical world is never to be finally solved is that natural processes waste all resources with an extravagance that man cannot afford. Man achieves his own purposes by those engineering interventions that check natural wastefulness in behalf of individual growth. All of culture, in our use of the term, is a damming up of natural wasteful flows—of heat and light and matter and seeds and time. Better civilizations will be better because they will not waste the powers of human beings; better civilizations will put more citizens to work at the most difficult things they can do. A democracy of culture would give every citizen, according to his capacities, his chance to make as well as enjoy what he lived by, and in this, also, human growth would be the measure of democratic reality. Every man would be doing whatever evoked from him the greatest congenial effort and gave him thus his chance to be his own best self.

WHAT IS SCIENCE?

I

WE HAVE STATED THAT OUR PURPOSE IN THIS ESSAY IS TO DISCUSS the use of science and the scientific method in the management of human affairs with freedom as the goal of the work to be done. We have said something of freedom. We must say something now of science. The method of freedom is the method of science; but that, although true, would not help much toward precision in fixing our terms. As we move ahead in a discussion of this kind, we try more and more to use words for the communication of exact thinking. Definitions are necessary. They will all be given in the way of science, that is, in the imperative mode. It is interesting to note that one basic difference between science and other ways of thought is that in literature generally, in poetry and philosophy and theology, definitions are in the indicative mode and most assertions of fact are in the imperative. Fact is absolute to the poet or the philosopher and descriptions of fact are compulsive, even if the compulsiveness is no more than a trace of what Morris calls the pragmatical meaning.[1]

In science, definitions are imperative after the manner of "let this word or symbol stand for this reference." Other statements are tentative or operational, that is, they are indicative but not absolute and not compulsive. The scientist, if he is philosophically well trained and knows what he is doing, says, "We use these words in these univocal and prescribed ways and we undertake to act on the generalizations stated. By this we can tell you what will happen if you behave in any one of the choices known to be open to you."

[1] C. W. Morris, *Foundations of the Theory of Signs* (Univ. of Chicago Press, 1938).

The other kind of statement is that such and such a word *really means* so and so, and in this there are traces of a false Platonic realism, a belief that the reality of the world lies in the symbols used to describe or point to it. From this use of words as absolutes comes the hortatory picture of the world which tells us not what will happen if we choose this or that, but what we "ought" to choose, regardless of humanly perceived consequences.

Here, let the word "science" indicate a body of knowledge and a method of apprehending experience. The knowledge comprises all description that is without adjectives of value except the adjectives "true" and "false." It is made up of abstract generalizations of observable events that are sufficient for prediction of future events in the same order. The scientist begins by defining entities (in imperative terms) and ends by measuring them. And the scientific method comprises all the ways of experiencing phenomena that will increase this kind of knowledge.

It is possible, without satiric intent, to say that the scientist, given a series of events, will make general statements that are sufficient to describe them and will not, with a pious bow to William of Ockham, go beyond that. The philosopher and the poet will usually make all the possible statements about an experience that are not easily seen to be incompatible with it. This way of dealing with experience serves the purposes of the poet or the philosopher because he is making out of events those emotionally charged patterns that give us purpose and ultimate design.

The scientist, with his different procedure, is trying to make a different kind of a pattern, one that is free of human purpose, except as it is a catalogue of opportunities. It appears to have been an error of Hegel to suppose that the thinking of the poet and the philosopher were actually determinative not only of human "truth," but of the world of nature outside man. This is exactly the opposite of the modern scientist's naïve nominalism in accord with which he "explains" nothing and offers mastery only by way of working descriptions.

The entities with which science deals in its advanced stages, however, are ideal or imaginary. Scientific generalizations are the abstract summaries of facts observable in the concrete or individual cases *in so far* as the actual cases approach the type. For example, the mathematician, who is supposed to deal in the purest of scientific materials, defines an entity in experience as a triangle, and then by the customary uses of observation and logic he makes generalizations about triangles, and measures aspects of their behavior. The fact is, however (*vide* Karl Pearson) that there is no such thing as a "real" example of the type to be seen, or touched, or smelled, or tasted, or heard. The type does not, in that sense, exist. All things subject to sensory observation, all things that "exist" are approximations to the type.

This is very important when we attempt to apply scientific methods to thinking about human beings. To remember it will save us from the common error of objecting to scientific statements about human behavior on the ground that they are abstract, as if scientific description could be anything else. For example, much misspent eloquence has gone into denouncing the economists for the use of the concept "economic man," on the grounds that there is no such "thing" as an economic man and that man never acts from exclusively economic motives. The proper answer is, of course, that no philosophically aware economist ever supposed for a moment that his economic man was anything but an ideal construct, defined arbitrarily for purposes of study. Whether it is possible or not to see the economic aspects of man's behavior in quantitatively measurable terms, as an independent variable, is a question to be answered by careful experimentation, and the quality of "economicness" is parallel, for scientific investigation, to the "triangularity" studied by the mathematician.

The method of science, as described here, should be seen in practice, not in polemics, and all this is by way of warning. We are going to discuss freedom as a value for which scientific mastery can be instrumentally used. As a matter of sociological observation we can say that science flourishes in a free social order, but we are concerned rather to show that freedom is attainable by scientific methods.

We might properly be surprised by the question, Can science in any substantial measure solve the problems of man's behavior? We have climbed from the first stages of humanity by taking thought and pains. Progress out of caves into cities, from simple cunning to today's machines, has not been by accident, nor by the working of any blind destiny, unless indeed destiny is only another name for labor and human thought. In dealing with their own natures, however, and with each other, men have been notoriously slow to learn and timid in using what they knew.

Since Socrates at least, in Western history, there have been some who have thought intelligence could be applied directly to the affairs of man, as it had been applied even before his time for many centuries to the machines of existence, to "things" and what we call the "material" world. Socrates believed in using the mind in human dealing and set a basic pattern for it in his search for the honest meanings of brave-sounding words. Social philosophers since then have repeated the expression of faith and a later line of thinkers have tried to put down in natural "laws" what is to be expected in the behavior of men. It would be difficult for anyone to say that they have been taken seriously as makers of single systems. Certainly Comte's sociology and the others like it are not accepted by all sociologists as abstract descriptive science of social behavior. Men continue to go on saying defiantly that we are going to use our minds to correct our own human errors, but in passionate disregard of such reasonableness, most men still try to solve all their serious problems by the ancient pattern of fairy tales, dreams, and violence that has always failed them heretofore.

The questions of whether or not what is known as science can alone solve human problems and whether or not it can solve them all are not issues now. Nor is this a claim that science can or should determine ultimate values. We began with an assumption of an ultimate value, freedom. We have told what we meant by the word freedom. Now we are looking at scientific ways of thought as the tools for the achievement of value and no accusations of "scientism" are in order. My only dogma is that we can usefully try the scientific method to find out how far science can help us in this unscientific purpose.

Man is a complex being and every man is unique. True. So is a bar of steel a complex being as anyone knows who has followed the growth of that great product of the imagination, molecular physics. So also is every bar of steel different in subtle ways from every other. But neither the complexity in molecular structure, nor the irrelevant particular differences in any bar, will keep an engineer from examining it by the general rules of the metal's strength, nor from fitting it into his bridge or his building where he needs it. He builds a steel frame, not by the particular qualities of his units of material, but by their general qualities, abstractly described. He builds by means of known general qualities that suffice for his purpose.

The example is directly relevant. Men can make progress toward building the kind of a society they want by making use of abstract, generalized, and often numerically expressed scientific facts about persons and institutions. Life can only be lived and enjoyed by individuals, by unique persons. But the social and cultural structures by which a good life can be made possible will be more swiftly and surely built if some of our thinkers and workers deal with relevant general ideas and forget that each man is unique.

An equally cogent analogy can be found in the science of physiology as applied in the practice of medicine. The good physician treats the whole patient; we did not need the Gestaltists to tell us that. And he thinks of the patient as a unique being. But when the unique human being has an earache the doctor makes use of highly generalized and abstract descriptions of earaches at large, depersonalized earaches. It does not matter who had them except in quantitative terms and this man's concrete pain is relieved by the abstract and general statement of what was learned from the pain of others.

There is much confusion on these questions. For our immediate purposes it may be enough to point out again that science could not be of use to anyone if it were merely descriptive of the concrete, single case, just as an experience that happens in only one life cannot be a rule for others. Only what can be generalized is socially useful. So we say that we ought to approach the problem of human freedom by discovering what generalized descriptions of human behavior can help us to build a free society that will work. The claim to be dealing in science is a high claim, difficult to make good. But we ought to be as scientific as possible, and when we make a judgment on the basis of a wish, it should be labeled. With the fairy tales and violence, the other parts of the old pattern, we need have nothing to do.

Most of the comments on the possibility of a social science that is like the physical sciences in its data and its statements of uniformity beg the main question which is this, Would the statements describing uniformities in social behavior, if they could be made, involve the phenomena of awareness in any way different from the involvements that are inescapable in other sciences? The only answer to this question is, of course, to try to find out.

This bears on our more specific question, the possibility of scientific human engineering, or scientific social reconstruction in natural sequence. Can there be precise description of human behavior, adequate for statistical prediction? Can we make, in engineering, applications of what we can know? It is always difficult to say which faction stands most in the way of the experiments that would give us answers to these questions, those who merely deny the possibility of social science in this sense, or those who cry for applications of half-achieved description. At the moment, we have to do with the too patient intellectuals, not with the impatient moralists.

II

While the struggle goes on to establish operational ways of thinking about social facts, we are losing, in our other scientific thinking, the mechanical idea of the nature of reality. That the world is a machine may be just now dawning upon some peripheral minds; indeed it may be a new element of healthy rationality to many persons whose immediate cultural forebears were steeped in animism. But it has been abandoned, not always with clear self-consciousness, by the most important scientific thinkers, and it will in time be less congenial to the rest of us. We may have soon a generation of men who will have lost, by the slow filtering down of the higher scientific generalizations, into the unconscious postulates of the general public, the whole mechanistic idea of reality. Will they be more willing to think scientifically about human behavior than we are now—or less so?

It is possible, perhaps, that this change in general thinking will never come, although modern physicists have mostly lost the idea that existential reality is entirely mechanical in its nature, that the world is a machine. They no longer believe that you cannot believe in the reality of anything unless you can make a model of it. They say they have given up thinking of a Creator who is an architect to substitute a Creator who is a mathematician. But to an outsider they seem to have made this change with little attention to the implications of what they were doing and with almost no change in their thinking about subjects other than physics. Apparently some of them can believe that an equation by Einstein or Broglie is an operational description of the material universe, without changing their other general ideas, even about the material universe. If that is true of the highly trained and active minds of these gifted men, general thinking may not be much reformed. The textbooks and popularizations by which the rest of us are supposed to learn something of material reality will not, for a long time to come, take the trouble to call attention to the fact that the Newtonian world, the world as machine, is not now being looked at nor thought about by the physicist. Each phase, as Einstein himself has pointed out, always includes the others and mechanical reality is included in the relativistic real.[2] But the addition of the new element gives a new basis for concepts. When will these new concepts be the basis for general thinking?

If the change does come, its whole effect should be good. Those who say that scientific methods useful in the investigation of the material world cannot be used to describe "social" phenomena, which are by our definition also part of the material world, appear to think that physicists still work with architectural ideas and that nothing that cannot be modeled can be conceived or believed in. Social behavior patterns are obviously not the sort of thing that anyone can put into a convincing three dimensional model. They exist in the mind and they show themselves in the uniformities of overt action. They exist, but not, perhaps, as the universe of Newton existed. Hence—so goes the argument with this postulate unconsciously introduced—hence they cannot be studied, defined, measured, and reduced to general "laws." If we lose our dependence on the Newtonian picture of "reality," we may possibly be less concerned about the "reality" of social entities. Incidental misinterpretations will doubtless plague us. The common fallacious notion that the Heisenberg principle of uncertainty indicates that the scientist now studies a "willful" or a "disorderly" universe is a sample of what we may encounter. But the faith of science is that more knowledge always helps, and this applies as well to the social group as to the individual. The relativistic world of nature may be the world in which man will effectively study himself.

[2] Albert Einstein and L. Infeld, *The Evolution of Physics* (New York, Simon and Schuster, 1938).

THE NATURE OF SOCIAL CHANGE

I

WE SAY, AS A DECLARATION OF OUR VALUES, THAT FREEDOM IS WORTH the vigilant effort needed to get and keep it. This is not the same as admitting that there are no reasons behind the declaration of such a judgment of value. We are not worshiping a word. Nor are we worshiping an unexamined way of life. We believe in freedom because, among other reasons, rational change is less costly in a free society. We are approaching the study of social engineering, and the purpose of social engineering would be the rational control of social change. We can thus show an affinity between our postulated end and the means now being discussed.

If we did not see change all around us, we should still know that it was inevitable by the reading of history. Very simple societies have kept their forms without much alteration through long periods, but they too change in time and no industrial culture such as ours can stand still. At least, none ever has, and there is no evidence on which to base a plan for a stationary social scheme. To discover the balance between change and stability is a problem of first importance; social engineers will have to deal with it. But some change there will be in a world like that of western Europe and America even if nothing happens but the going to pieces by obsolescence of our present industrial scheme. Invention may fail, but there will still be shifts in the habits of men as in every other observable phase of the natural world. Freedom is the condition in which that change can be achieved with the most rational good will and the least damage.

The heirs of great fortunes care very little for the pains that went into the making of the goods they enjoy merely by being born, and in much the same way, every generation of men, never thinking of itself as "posterity," accepts without thanks the benefit of past change. As the heirs of worldly goods are likely to underestimate their own good luck, so succeeding generations make very little of the human suffering that brought them their commonplace advantages. The record shows that changes, up to now, have cost too much; not more than they were worth, but generally much more than they needed to cost. In many cases, in so far as one can reconstruct history in the imagination, it is safe to say that the good results of revolutions and wars and other social catastrophes could have been secured even more quickly by other means. What we need to know much more about, what we need to investigate with the most rigorous and scientific method, is the nature of those other means.

Men are incurably poets, and even the historians who pretend to make sober judgments on the past are naturally seduced by the drama of dead wars. Social change that was accomplished without heroics or blood is not exciting in the telling and historians are as easily bored as other men. So we know a good deal about all past change that was spectacular and about all the spectacular phases of the duller years. Men have often, perhaps nearly always, been more miserable in the past than the historian can picture them but almost never have they been so dramatic. It is a favorite idea of gentle satirists, like Anatole France for example, that those who have lived through great crises scarcely knew what was happening. In some degree, this is because what really happened, *wie es eigentlich gewesen ist,* was just this day-by-day ordinariness. The perspective of the chronicler that enables him to make it seem so much more dramatic is like the selective faculty of the landscape painter: It lets him leave out of the natural scene all that dulls his sharp, contrasting, dramatic effect.

The sociologist, looking not for drama but for uniformities in human behavior, can see that change is often quite without violence; violence often fails. The picture of the past that the social scientist creates with the same materials is more monotonous and more quiet. But the dramatic scenes that were actually lived by someone else, someone now dead and out of misery, the wars and revolutions and catastrophes, are much more pleasant in retrospect than they were to live through. It may quite well be that science will make social change some day so quiet and efficient that life will look dull to all except those who are living it, dull, that is, to the historians and the poets of posterity. No happier achievement could be hoped for!

II

Social change, or change in collective ways of behavior, can come about when persons who have social power change their own behavior, when persons with different social habits come into power, or when persons of one set of habits die and are replaced by persons of habits that are different. A mechanical change like an invention, a major shift in circumstances, or even a catastrophe such as war or famine, will not cause social change unless adaptations are made to the new conditions. Inventions can be made as mere mechanical contrivances and still not be put to any use and, therefore, be meaningless in culture. Tribes have chosen to perish rather than make adaptations to catastrophic shifts in the environment. Change in society is change inside the persons of an important group of the society's members, or else it is a change in *dramatis personae*. It can happen only inside of persons; circumstance can be its instigation or its material tool.

There is a steady and, in most societies, a very slow kind of change that comes because culture must be learned. Each generation makes its own version of essential behavior forms. This is something more than the constant shifting in manners that causes irritation between every set of parents and their children. The family conflict, even in very static cultures, seems unavoidable because the old and the young do not look at life in the same way. Old eyes look backward and the young look ahead and no landscape ever was the same from two different points of view. This irritation is not without social importance. It has been suggested that the very first social institutions may have been devised to adjust the quarreling between fathers and sons. Much of that quarreling is about different versions of the same culture, but underneath what the passing generations see and argue about are slow drifts of a more important sort. Because all men die and their ideas can be handed down only by way of the mental grasp of them that each succeeding group of heirs can get, faulty learning is a factor in the remaking of the world.

We can see this in an institution basic to all culture, language. Linguists[1] say it is quite unlikely that any generation of children ever pronounced or used any language exactly as its parents did. And in a century changes are very great.

It is interesting to speculate on Shakespeare's bewilderment if he could come back and hear a modern actor doing Lear. And we may soon begin to speculate on the probable effect on language change of modern methods of recording sound. The radio evidently tends to smooth out regional differences in speech. Will the phonograph freeze linguistic forms? If we can sometime know more of these things we shall know more of this phase of social change, the kind of change that comes from imperfect learning. But it is probably not the most important phase that we have to consider.

A swifter and more dramatic change, also by reason of new *dramatis personae*, new bearers of the culture, comes when a social or political revolution puts new persons into power. These are the changes affected by sheer coercion. They may not represent any inner shifts at all except a bow to force. In pure form they are very rare.

Most new patterns come in the collectivity because individual persons who make the group are themselves changed—internally, that is, in their pattern of predictable behavior. The person is definitely modified in some essential way and enough of such transformed persons exist together in the social group to change

[1] J. Vendryes, *Language*, tr. P. Radin (New York, Knopf 1925), p. 40; A. H. Sayce, *Introduction to the Science of Language* (London, Kegan, Trench, Trübner, 1890), I, 169; Louis H. Gray, *Foundations of Language* (New York, Macmillan, 1939), p. 85.

the institutions or their proportionate balance of social power. The locus of social change is in persons, and we look at persons to watch it happen.

The statement that social change generally comes when individuals take on new patterns of behavior is, at first thought, a too obvious observation. But much of the difficulty that lies in the way of explaining social change is in the failure to keep this too obvious fact in mind. Most readers of sociological literature are familiar with an experience that goes something like this: They begin to read a treatise in which it is written, in precise terms, that the chief problem of social analysis is the description of the processes of social change. Or they begin a less rigorous work which announces with betraying assurance that all social change is the result of this or that one element in the world. In both cases, as the argument goes on, the reader is baffled to find exactly where it is that the phenomenon takes place. It would appear that the first step in analysis might well be to locate the events to be analyzed. The assertion that they take place in the personality pattern of individuals is, therefore, not too commonplace to be urged. To list the factors such as invention, environment, or even transcendental ideas, is to provide description of the energies at work. We still need to know what material they work on, where they are organized, if we hope to observe the changes taking place.

Social change happens aside from revolutionary shift of those in power, and from the succession of careless heirs, when a significant number of the members of a social group either change a specific loyalty, or change the proportionate strength of two related loyalties, so that one which had been dominant has now a lesser claim. A new dominance always succeeds.

The greatest changes in institutions may be buried away from our sight in the unwritten history of cultures, but there are events that can be studied in the records. Some institutions of the cultures of the Western world have ceased to exist after having been taken for granted for centuries. Others are dying as we watch. Slavery, for example, not wage servitude or other forms of cruel injustice but the simply and easily defined chattel ownership of the body of one human being by another, is gone.

Caste is dying away.

War is in serious question.

These three, slavery, caste, and war, have been fundamental cultural structures; they have been necessary to all kinds of related institutions. We can make logical guesses at their origins; but it is more important that we can observe their changes and perhaps get from that observation the principles by which we can hurry other changes in the direction we want them to take.

Slavery is supposed to have begun when the first conqueror spared the life of a defeated enemy and took him home. If that is what happened, slavery was a step in progress, since it gave life, of a sort, as a substitute for quick despatch. In behalf of the masters, it provided a steady labor supply for cultures that were struggling to become industrial. It comes under our direct examination in Greece where the household economics would have been difficult and the mines probably impossible without expendable human energy.

Aristotle said, "He then is by nature formed a slave, who is fitted to become the chattel of another person. . . . It is clear then that some men are free by nature, and others are slaves, and that in the case of the latter the lot of slavery is both advantageous and just."[2] This opinion was held with hot sincerity by many American Christians in 1860 and a few of them are possibly still alive. And Aristotle, like a Georgia planter, could show what he thought was proof that the economy of his culture required slave labor. Aristotle's countrymen, especially in the period following Macedonian expansion, had a factory system of considerable complexity and "Greek influence" in the Mediterranean was in the direction of an industrial revolution.[3]

[2] *Politics*, tr. Walford (London, Bell, 1880), I, v.

[3] W. L. Westermann, *Upon Slavery in Ptolemaic Egypt* (New York, Columbia Univ. Press, 1929), p. 55.

In the plantations of America, a primitive agriculture wasted labor as it wasted the soil. The technological changes of the middle nineteenth century were making slave labor less and less satisfactory, and it may be that even if there had been no Abolitionists and no Civil War, slavery in the United States would have been at last abandoned because it was unprofitable. It had been unprofitable in New England for a long time and had not been much practiced since colonial days, so the New England reformers had no need to struggle with a desire for profits. So it was in Britain. British shipowners were willing to seize Negroes in savage Africa, and sell those that survived the voyage in American markets, but white factory hands of all ages could be bought in their own country for subsistence wages without importing blacks. So there was no slavery in the British Isles.[4]

We have no means of knowing when these practical reasons would by their own strength have ended slavery as an institution in Western culture. In the meantime, there was an idea loose in the world with which chattel slavery was incompatible. Whitehead has remarked on the length of time it took Plato's assertion of the value of the unique human soul to be transformed into the abolition of chattel slavery.[5] Second and third century Christians would certainly have thought slavery incompatible with their own religion and since many of them were slaves they saw liberation (in Paradise) as a logical end to their earthly faithfulness. And they also practiced freedom until Christianity triumphed by being adopted as the religion of power and took on the sins along with the mastery of its secular world. The moral protest went on through all the ages. No one can say in this or in any other cultural transformation what may be the relative importance of mixed social motives. Some men work for humane causes because they love to help humanity; some men

[4] C. M. MacInnes, *England and Slavery* (Bristol, Eng., Arrowsmith, 1934).

[5] A. N. Whitehead, *Adventures of Ideas* (New York, Macmillan, 1935), pp. 15 *et seq.*

accept reform when they no longer think the selfish advantages of an older system worth hanging on to.

Slavery ended in America, its last home in the civilized world, when moral revulsion against cruelty was combined with economic foresight and the catastrophe of a war. It had ended in European countries much earlier and for less clearly seen reasons, but there, too, combinations of advantages and ethics played a part. In our standard of what is cruel and what is humane we are as inconsistent as in other ethical judgments.[6] All steps forward appear to be combinations of these circumstances: expediency and principle; and it would be hard to find a great social change in which either one was lacking. We have no need to dole out credit. We are concerned to determine what combinations of moral and practical reasons can be found to support future progress.

III

More can be learned from the study of a not yet dead but dying institution, caste. Its history and present condition can only be touched on here; some of the energy that now goes into unintelligent, because undirected, reform might well go into an examination of the reasons for the decline that has already taken place. And as slavery means chattel ownership of the body and not mere cruel oppression, so caste means a fixed social position by reason of birth and not mere "class" difference. An individual who suffers disadvantages from a menial position or lowly job is not a victim of caste in our sense unless his birth compelled him to take such a place and denied him any chance at another, no matter how great his natural powers. Slavery systems and caste systems have much in common.

A caste system is almost intact in modern India. The British *raj* has suppressed some of its more severe cruelties but gets little credit for this humaneness because the British are a new caste

[6] W. Stanley Jevons, *Methods of Social Reform* (London, Macmillan, 1904), p. 209.

in themselves. And because Indian culture is more primitive than Western industrial civilization, the caste system is entwined with religious, social, and political institutions and affects them all.

We can believe the tradition that the beginnings of caste in India were set up by conquerors from the north who came among the natives of lower culture and different race. The Aryans, the shadowy white folk who spoke the primitive form of modern Western languages, were superior to the dark savages and the simpler black folk, and they made that superiority lasting by fixing the racial groups in caste ranking.[7] Religious and occupational distinctions followed. And it may have been a wise adjustment of the contact between the native and the northern hordes, just as slavery was a progressive invention.

In biological thinking we need always to remember Aristotle's caution not to judge anything by its origin but by its present value, since biological evolution has been, in general, from the simple to the complex. But in the history of social change we have to exercise a tolerance in reverse, and not judge the value of an institution in its beginning by what we see of its operation now. Much zeal and indignation is wasted in regretting that modern values did not exist before there came into being the conditions that made them possible. Caste is wholly cruel in its extreme degree in our Western eyes, but it once may have been kind.

In fact, we practice caste in our own society. In the European and British phases of Western culture, there is a smell of caste in the educational and social provisions for different strata of the social group which make birth an almost insurmountable bar to some kinds of personal promotion. But the grading is not absolutely rigid. In the United States of America, the Negro lives

[7] V. Gordon Childe, *The Aryans* (New York, Knopf, 1926), p. 159: "In Indian literature the word for caste is *varna*, 'colour,' already in Vedic times, and since the Pandavas in the later epics are described as tall and fair, it may be concluded that the distinction intended was that between swarthy Dravidians or pre-Dravidians, *dasyus*, described even in the Rigveda as black-skinned (krisnavarna), and fair Aryans." (Reprinted by permission of the publisher.)

in a caste system. Brought out of slavery suddenly and without the consent of his masters, the Negro is born to disability. The social and economic motives that support this caste structure are not hidden.

In these transformations, ridding us of slavery, mitigating caste, and perhaps in the future ending war, there is a movement analogous to and partly dependent on the movement which Boas has described as the decisive trait in the advancement of civilization. The ethical radius, which describes the in-group with whom we are at peace although all the rest of mankind is enemy, has been constantly lengthening.

> Thus the history of mankind shows us the spectacle of the grouping of man in more or less firmly knit units of ever-increasing size that live together in peace, and that are ready to go to war only with other groups outside of their own limits. Notwithstanding all temporary revolutions and the shattering of larger units for the time being, the progress in the direction of recognition of common interests in larger groups, and consequent political federation has been so regular and so marked that we must needs conclude that the tendencies which have swayed this development in the past will govern our history in the future.[8]

This slow change affects the relations of all closed groups with persons outside and also weakens progressively all disabilities inside the group that are caused by status rather than by conduct.

Great institutions, like caste, founded on obscurely righteous attempts to meet practical problems, have too much strength to be easily destroyed. It is not only that better and better reasons are often invented for institutional habits as their original reasons for being disappear. Rationalizing of habit is a strong conserving

[8] Franz Boas, *Anthropology and Modern Life* (rev. ed., New York, Norton, 1932), p. 101. (Reprinted by permission of the publisher.) Or, as an early American philosopher guessed in 1799, "It seems then that the tendency of civilization is to diminish the number of nations." Joel Barlow, in his pamphlet "To His Fellow Citizens of the United States," quoted in Charles Beard, *American Spirit* (New York, Macmillan, 1942), p. 141.

force but the mere size of the institutionalized group is also a factor in the changing situation. The tendency of men in civilization to collect in larger and larger groups, in which Boas rightly sees the most beneficent of all the effects of gregariousness, can sometimes work temporarily to protect from change the institutions that have lost their right to exist.

An example may be cited from an observation in India. As a traveler in India in 1922, I had a "bearer," a servant, who acted as porter, valet, guide and (suspicious) friend. He was a Moslem by dress and eating habits and by fanatic affirmation. He also had innumerable relatives in accord with the fabled fecundity of the East, and one of them died just in time to take him from me for a day or two whenever we arrived in a new city. His proud Islamic ancestry would have been shocked to see how sly his condition of servitude had made him. But more shocking was his profession to me of his "caste" position. I tried in academic ignorance to remind him that Islam was the most democratic of religious communities. He was living in hostile nearness to caste-proud Hindus and he was by reflection living in a caste system which had evidently been made by the hardening of sectarian and occupational differences into status lines.

In such a case, the great size of the Hindu community, organized in rigid caste differences, had completely changed the invading egalitarian Moslems. The smaller group, although now many millions, is infected with the traits of the larger group. This is regression, a historical accident. The Chinese, who make up that other great Eastern bloc of peoples, are altogether free of caste.

Caste, in India, dies first in the minds and convictions of the most highly developed individualities in the culture. And the same thing is true in the slow deterioration of the caste system in America. It would be in the school of Rabindranath Tagore that one would expect to see caste differences ignored. It would be there, also, that one could see ignored the institution that

Islam, some say[9] *en revanche*, has fastened on the Hindus, the sex-caste, the subjection of women.

Caste and sex equality would be possible to a great saint like Tagore but more difficult for a great nationalist politician like Mahatma Ghandi. Ghandi has long been opposed to the more cruel practices of "untouchability," but as an orthodox Hindu he holds to the basic idea of exclusive hereditary groups.[10] And what one could see at Bholpur when Tagore's spirit ruled there can be seen anywhere in lesser degree. The persons who are at the top of cultural development can ignore caste, as when a Negro scholar is treated without difference by his scholarly colleagues, or a Negro statesman by a Northern government official. Caste dies first in the minds of those who are most "enlightened," because enlightenment not only makes a person more imaginative and sensitive but also breaks down somewhat the instinct-like fixity of group behavior.

These palliations cannot be depended on, any more than kindly masters could destroy the slavery they softened, but it is, on the other hand, very unlikely that kind masters of slaves did in fact conserve the evil by making its impact less decisively cruel. Their kindness is a symptom of that general condition in which humaneness and interest can work together rather than against each other and in which social good has the best chance of being accomplished, orthodox Marxian doctrine to the con-

[9] The beginnings of that degradation of women go far back, however, probably to the time when Manu's legislative precepts put a rigid formality on the restraint that was destroying the pastoral freedoms of Vedic times. See Clarisse Bader, *Women in Ancient India*, tr. Mary E. R. Martin (London, Trübner, 1925); and see also the Laws of Manu. Apologists for Hinduism rely too much on the primitive or Vedic freedom of women. The edict of Manu seems to have put women in an inferior position in India more than a thousand years before Islam began. Even the Moslem household, while it was still primitive in spite of polygyny, gave women a good deal of respect. The degradation of women in India was probably much more complex and much more due to cultural catastrophes than doctrinal or communal pride would admit. The relation of sex-oppression to caste-oppression should be carefully studied.

[10] Mahatma Ghandi, *Young India, 1924-1926* (New York, Viking, 1927), especially pp. 653-57.

trary notwithstanding. Caste will disappear for the Negro, perhaps even in the far future for the Hindu in India, when man can no longer profit by it either spiritually or materially.

Against any such institution, the decisive weapon of those who work for cultural change must be forged of knowledge itself. And this knowledge must be, in part at least, the knowledge that we have been talking about, the operational but universally valid knowledge of science. I say "in part at least" because Christian charity, based on credal absolutes, has softened the cruelties and weakened the structure of Indian caste. The Indian peasant, usually the untouchable or outcaste, frees himself altogether from the bonds of caste when he enters a Christian community; and Christianity in India, as in its first days in the Roman Empire, grows mostly by giving a not intolerable way of life to the oppressed. It saps the cruel fixities of a caste culture at the base of the pyramid of power. Christianity may even supply the sufficient motive for the complete destruction of caste in India, although one may be allowed some skepticism. Few missionaries are working at home to destroy the caste of the American Negro.

The material and the objective truths must work with the charitable motive before caste follows slavery, first into disuse, and then into the realm of the unimaginable. India will become in some degree urbanized and industrialized, although probably not in the degree we know in Western culture. Secular education will weaken custom and Hinduism itself may lose hold. It may be difficult for those who have ever lived for long among Hindus to imagine a time when it will not be a virtue to inflict bitter oppression on an "untouchable" because his birth is clear evidence of Heaven's displeasure. That is an almost impossible feat of imagination for the typical Anglo-Indian, just as it is almost impossible for a kind-hearted Southern American of white blood to imagine a Negro at his table instead of behind his chair. But neither change is difficult for the Frenchman or the Scandinavian who has had little experience with caste. As I have said, most real progress in any set of institutions must be in changes that are not only beyond the imaginations but outside the desires of the typical end-products of the progress of the past.

IV

When an evidently evil institution is still strong in our habits and we see examples of it in contiguous cultures, we are likely to call it "human nature" and thus try to escape responsibility for correction. With something of the same irony that makes us officially call anything an "act of God" when it is catastrophic beyond human invention, whereas we generally credit no deity with our steady privilege of existence, so we call "human nature" whatever seems incurable and incurably bad. There are not many men left now who would call slavery "natural," and there are few who would still apologize for the dissolving frames of caste in our society on the ground that they were human nature. But war is in human nature, we are told, and therein are mixed up, by a confused and cowardly evasion, all the false ideas of both war and human naturalness. The cowardice, the confusion, and the whole evasive process are also "natural" and should so be understood.

The social engineer, the scientific humanist, if he is convinced that war is evil, remembers that it is an institution, a shared habit like any other, rational in its beginnings perhaps, but no longer useful. Out of his observations of the more or less rational decline of other institutions, he tries to discover principles by which he can move to end it. All things that he can discover about the human nature underlying institutional behavior are data, that is, neither obstacle nor help, since he is working by the principle that we gain control only by understanding the natural facts as described in scientific laws, not by ethical judgments or any other judgments of preference.

It is quite possible, indeed it is being done all the time in our culture and has been done in much simpler societies, to subdue the emotional drives of men to acts that appear to be quite contrary to the natural gradients of energy. That is to say, man can learn to behave in such a variety of ways that we almost believe him to have sophisticated or obscured in himself whatever there is of original nature. Human nature in culture is capable of almost unlimited varieties of expression for the same drives.

Consequently the social engineer seeks in such a problem as the abolition of war two things, the scientific ways of describing war that will provide the abstractions that can be manipulated to indicate effective action, and the relevant scientific facts about persons. We have been indicating the qualities of the first set of facts in our discussion of slavery and caste. Further discussion here is only for exemplary purposes and cannot be complete. It has to be said, however, that most modern studies of war, even scholarly works like Quincy Wright's[11] are mostly historical rather than sociological. The compilation, ***War as a Social Institution***[12] is full of useful hints. But so far as I know, no one has examined directly the development and decline of comparable institutions in order to find the necessary mode of operation. Almost no one has looked at the problem with the engineer's direct and practical intention. Historians and sociologists in other ways have tried to find the theoretically useful scientific generalizations and the unique facts. Statesmen—and one hopes to avoid being unjust to these much maligned servants of our national lusts and unacknowledged pride—appear to be concerned with avoiding the "next war" rather than with changing social patterns. Applied anthropology in government, as seen, for example, in the Department of Agriculture's operations directed by M. L. Wilson, are an example to the contrary, but it should be noted that these operations are not the work of elected officials nor much dependent upon the understanding of a legislative committee. The anthropologists, especially the group now boldly

[11] Quincy Wright, *A Study of War* (Univ. of Chicago Press, 1942).
[12] J. D. Clarkson and T. C. Cochran, eds. (New York, Columbia Univ. Press, 1941).

trying to make an applied science out of their knowledge, offer what appears to be the most hopeful of new moves.[13]

The needed "laws" of behavior in culture, relevant to the abolition of the institutions of war, are not yet available, at least not in any accessible form. They must be dug for. Many of them are still to be created by observant men, but we can see that they are in process. The social engineer, if such laws can be created, will need to examine the data of intrinsic or original human nature on the psychological side while he also finds out whatever he can about the institutional context. There are strong reasons to be found in modern psychology, especially in several schools of psychopathology, for believing that the war institution is a useful means for the discharge of personal aggressiveness which is in turn a form of energy created by the frustration of natural drives. A practicing psychiatrist calls the group behavior of modern Germany by the name of a personal psychosis "paranoia," [14] and another seeks the roots of our uses of the war institution in individual fears.[15] One may hesitate to follow these suggestions that may be oversimple, but we can certainly understand behavior better because of the insights of psychologists and anthropologists who have taken hints or leadership from Freud. Much may be learned from Bateson, Mead, Gorer, L. K. Frank, Lasswell, Kardiner, Benedict, and others [16] more or less of the same school. They do not, if I may lump crudely together the ideas of independent thinkers, believe that wars are either natural or necessary expressions of innate human qualities. They see in social patterns, in cultural matrices, the stresses that both frustrate individual powers and turn these powers to expression in the institutions of war.

[13] Cf. the publications of the Society for Applied Anthropology, ed. Eliot D. Chapple, Boston, Mass.

[14] Richard Brickner, *Is Germany Incurable?* (Philadelphia, Lippincott, 1943).

[15] Erich Fromm, *Escape from Freedom* (New York, Farrar and Rinehart, 1941).

[16] Cf. B. Malinowski, in *War as a Social Institution*, ed. Clarkson and Cochran (New York, Columbia Univ. Press, 1941).

As psychologists, we shall have to dig deeper still. There are relevant facts in the behavior patterns of socially adapted, nonhuman animals. The biologist has something to say about the roots of violence and cooperation in nonhuman societies. An authoritative comment by a student of animal societies on the post-Darwinian belief in the naturalness of ruthless competition is: "The biological support for this fatalistic view regarding, among other things, the inevitableness of intra-species human conflict, is now opposed by strong evidence which indicates that the idea of a ruthless struggle for existence is not the whole, or even the major contribution of current biology to social philosophy and social ethics." [17] Professor Allee says more summarily, "Despite many known appearances to the contrary, human altruistic drives are as firmly based on an animal ancestry as is man himself. Our tendencies toward goodness, such as they are, are as innate as our tendencies toward intelligence; we could do well with more of both."

Above all, the engineer will work by taking "nature" into account, rather than by any attempt to change it. The chemist does not work miracles in the plastic reconstitution of molecules by "changing" the "nature" of the chemical elements. And the social engineer will take into consideration all that he can observe in the institutional context of the time in which he works. It may be true, as was said by Theodore C. Hume [18] that men will fulfill their "freedom" by fighting wars as long as they want above all else to be strong and rich. The possible answer to this is the one made fruitlessly by Norman Angell, before the first World War, in *The Great Illusion,* when he predicted that wars under modern industrial conditions would never bring prosperity even to the victors. He appeared then to prove, and still appears to have proved his thesis, but he did not convince the ruling groups in several of the most intelligent and culturally developed

[17] W. C. Allee, "Where Angels Fear to Tread; a Contribution from General Sociology to Human Ethics," *Science,* XCVII, No. 2528 (June 11, 1943), 517. Reprinted by permission of the publisher.

[18] In a personal conversation with the author.

of modern nations. Both the Japanese and the Germans set out to get material prosperity by way of war.

The engineer will need also to take into account some of the problems of world scope that make for violence because of the cluster of institutions that are comprised in "nationalism" and "national sovereignty." For example, since differences of ideals can struggle for mastery without physical violence in a pluralistic society, why must they be reasons for war if they are international? Religions, political faiths, even sharp contrasts in basic ideals, can live in competition in the United States. They have been doing so for a century and a half. The mechanisms of the sovereign national state are required to make these same differences of opinion into reasons for mutual destruction. It is a fair question: Can we reconcile deep differences in culture? If we cannot, can we nevertheless make international pluralism or diversity as quiet as diversity is in a free state?

On the material side, we have the problem of wide differences between levels of material prosperity in different peoples. Reformers, liberals, men of good will, have talked lately of making the whole world a better place for the common man. The purpose is admirable. But the simple assumption that good will can do the job is not a scientific nor a practical conclusion, nor is there indeed much scientific reason for believing that it can be done by any means now known. To put it briefly, since the whole question is distressing, we have not solved, even in this most prosperous United States of America, the problem of the gap between the income of farmers, miners, and woodmen, the primary producers, on the one hand, and, on the other hand, the incomes of factory workers protected by trade unions, and the clerks, the professional workers, and the merchants.[19]

The farmers of the United States make up, in normal years, about one-fifth of our working population. They are doing very well if they get one-tenth of our total income.

The same difference in possible material reward for labor as

[19] Cf. Colin Clark, *Economics of 1960* (London, Macmillan, 1942).

exists between the primary producers and the tertiary producers in the United States exists as a possibly permanent difference between industrial regions like western Europe or northern North America and the agricultural peasants of China and India, eastern Russia and South America, which make up more than half the population of the earth. The scholars who can speak with any real authority on the subject [20] tell us that these congested regions must, under present or now foreseen technological conditions, remain agricultural. They must continue to be primary producing centers. If we cannot, even with political and human pressures upon us, solve the problems of farmers at home, it appears to be sanguine beyond reason to think we can do it for 800,000,000 strangers. Can we be rich in the face of a poverty-angered world and be forever safe? The social engineer has human wills and human passions in his primary data and he must have an understanding of them in his primary principles.

V

The conditions of freedom for the person are the conditions that give full use to his creative powers. The same thing can be said, not by analogy but literally, of a society. The free society is the creative society. Whatever it may have in it of power to change, the power that must lie in the persons of its members will be brought in greater plenty into action according to its freedom. We have located social change in the conflict and shifts of loyalties in the person.

This will be an oversimplification perhaps, but we may speak of social change as causing the less suffering to human beings according to the size of the stage on which the conflict takes place. Catastrophic causes of change, the wars and famines and migrations, still bring their effects by shifts in the loyalty patterns of persons, but they work in patterns of conflict in which people are pitted against inordinate nature or they throw masses of

[20] C. K. Leith, *World Minerals and World Peace* (Washington, D. C., Brookings Institution, 1943).

people against people. In such cases global loyalties, such as patriotism and communal bonds, are overridden by the impulse to change, but many individuals die rather than change, and many more may change in shame and fear. In primitive conditions, change is accomplished on this scale by destroying the bearers of certain institutions. It must have happened often in the earliest and the simplest conflicts between tribe and tribe. It was the meaning of the Roman plow that turned in the ashes of Carthage.

Narrow the stage of conflict and you may expect some lessening of the cost. This is to say that violent revolutions in single countries cost less, on the whole, than wars, in proportion to the amount of change they bring about. There is, of course, an emotional intensity in conflict within a society, a hatred of brother against brother, that appears to be more venomous than hatred of the enemy outside. A sense of bitterness cannot be measured. When we say that revolutions[21] cost less for their results than wars, we can mean only that the observable destruction of material goods and social values appears to be less in revolutions.

The means of making such an observation are poor indeed, especially since most great revolutions follow wars and merge into wars. It is impossible to say what was cost of war and what was cost of social violence. And it follows that we cannot clearly extricate the good that has come out of the social violence from what has come out of war phases of the same upheaval. In the case of the most famous of all Western revolutions, the French eruption of 1789, there were wars preceding the popular revolt, and then came the Napoleonic phase and the European reaction, all flowing in the same nexus of trouble. What "caused" the balance on the credit side of this generation of slaughter? A material balance is to be found, probably, in the destruction of the oppressive landowning system that was grinding the peasants.

[21] Crane Brinton, *The Anatomy of Revolution* (New York, Norton, 1938).

But it was a slight balance, all things taken into account, unless we can add to that tangible change the more important change in the attitude of French leaders which made them undertake to apply to the lives of their subjects the enlightenment that had been preached nowhere so eloquently as in France itself. The cruelty of the *ancien régime* was a long drawn-out, exquisite cruelty, a dawdling, witty cruelty, whose brutality was scarcely evident in the behavior or temperaments of those who profited. The cruelty of the French mob, in revenge, was bestial and direct. The real gain, therefore, came in just those slowly accelerated measures that made the sufferers not only happier but less beastly.

The citizens of the United States of America are both parents and heirs of the French revolution. The example of their own ancestors encouraged the French revolt and their own democracy was encouraged afterward by the violent, brave therapy of the *sans-culottes*. This makes us poor judges of the significance and importance of that revolution. We have mostly forgotten the smell of blood that pervaded all talk of France at that time. The best and the innocent who were trundled to the street corners and murdered as recreation for the public were understandable symbols of past oppressions but they were still guiltless men and women, and it took strong stomachs, like those of Paine and Jefferson, not to be upset. Conservative, if not more humane men, Adams and Hamilton, for example, were afraid of the contagion. Even Burke, who had built on his defense of the Americans what Acton [22] called the "noblest political philosophy in the world," could not see liberty as good when it came at such a price. What Burke evidently failed to see was the fact that the bestiality of the mobs was the greatest indictment of the old rule. An obscure Delaware pamphleteer, Robert Coram,[23] answered Burke better than Paine did in saying, "Humanity is

[22] Lord Acton, *History of Freedom* (London, Macmillan, 1922), p. 56.
[23] Robert Coram, "Political Enquiries, 1791," in Charles Beard, *American Spirit* (New York, Macmillan, 1942), p. 126.

wounded by the outrages of the mob in France; but what better can be expected from *ignorance* the natural parent of all enormity." And John Morley [24] cut deeper, "If the majority of the people—that is, if the people—are ignorant and narrow, it is certain that any set of men who successively rule over them will also partake of that ignorance and narrowness."

All debaters in politics know that human sacrifice is needed in social change. The radical thinks it a small price for progress to have a few bourgeois heads roll in the dirt. His own head will not be among them. Also it is the material property of other men that will be confiscated. The radical seems brutal to the conservative who begs for peace. But peaceful change has its human sacrifices also, in the unrelieved miseries of the poor and dispossessed who are denied immediate help because changes too swift would upset the social structure. The members of society who have been successful in terms of things-as-they-are must be liquidated if change is too swift, and the defeated must expect nothing if change is too slow. It is not surprising that there should be recrimination between radicals and conservatives.

The social engineer who considers revolution as a possible method of getting a better order has to be above this debate, however, if he is to make an honest appraisal of the cost ahead. Revolutions are generally begun by men who are passionately resentful against cruelty and therefore not expert in violence. They are followed, in the pattern Brinton describes, by men who do not hate cruelty so much as they hate their enemies and so have no objection whatever to being cruelly violent for their own purposes. "Old revolutionists" are much like men who might be so pitying of a caged tiger that they would release him; there is always a look of understandable surprise on their faces when the tiger strikes at them as soon as he has finished with his keepers.

There is an intimacy and vividness in the civil wars we call revolutions that make it hard to tell whether or not they are

[24] John Morley, *Edmund Burke* (New York, Knopf, 1924), p. 249.

worth the suffering they cause.[25] What we want, what this argument is in behalf of, is a social order in which change can happen with the smallest possible degree of violence. The principle of such a society is this: The conflict among loyalties that brings on social change is narrowed down toward the stage of individual action. Social action is still group action, but the conflict involves only the strictly relevant institutions and so does not involve the whole person.

VI

We can make the principle somewhat more concrete by sketching roughly three possible stages in the progress of a society toward freedom. I will call these stages global, mosaic, and pluralistic. No society under actual observation ever fell strictly into any one of the classifications, but the islanders of Dobu described by Fortune[26] had something like a global society; many European groups (the Austrians, for example, before 1914) have had a mosaic society; and we are now, in the metropolitan centers of industrial America, approaching a pluralistic society.

The word global here is used to mean all-of-a-piece, simple and homogeneous. The global society, if one could be found that was perfect of its kind, would have a few easily seen characteristics. The pattern of loyalties in every person of the same age and sex in the group would be the same. The only differences would be based on the age and sex differences that are universal and so simple as to be recognized in the shared habits of the simplest groups. In such a society, the pattern of loyalties is the same in everyone because there are no institutional choices. There is basically no division of labor except in sex groups. In the fishing season, all able-bodied men fish. In the seal-hunting season, they all hunt seal. There are loose and transitory patterns of leadership and cooperation but the binding loyalty is to the

[25] Cf. Everett Dean Martin, *Farewell to Revolution* (New York, Norton, 1935).

[26] R. F. Fortune, *Sorcerers of Dobu* (London, Routledge, 1932).

group as a whole. The religious loyalty in the self-pattern of every person is to the same institution as it is in every other person because there is only one possible religious belief. The work community, the religious community, the political community, are all phases of the global community, the community as a whole. There are normally no rivalries or questions or choices in any of these fundamental relationships. We know now that primitive societies were not so changeless as they once seemed to be.[27] There is a slow erosion there in the process of cultural inheritance. Invention is not always punished as treason. But change itself has no institutional means; there is no prediction and little preparation for change. The unexpected is not expected as it is in societies of more complex systems.

The person in the primitive group, however, is made up of the same impulses and continuities as the person in a complex group; he goes mostly by habit but he can invent. The comparative slowness in change of the typical primitive group is not the result of totally suppressed inventive powers. Much of the savage's inventive power is drawn off into improvement instead of change. Marett [28] explains how the primitive uses his creative abilities to make better tools with which to hunt and fish, instead, for instance, of learning how to do something more economically efficient than fishing, like raising crops. He makes "progress" in the same categories of labor but makes no change. Thus his innovations can be real without strain and without drastically affecting his loyalties. There would always be some conflict; the inventor is always a social enemy in some measure because he assaults the respect for self of everyone who has established his personal habit-pattern on the current model. But an improvement is much less insulting than a change.

When thorough change does come to a global society, however, it comes almost necessarily as catastrophe, and only a

[27] Cf. R. H. Lowie, "The Transition of Civilization in Primitive Society," *Am. J. Soc.*, XLVII, No. 4 (Jan., 1942).

[28] R. R. Marett, *Faith, Hope and Charity in Primitive Religion* (New York, Macmillan, 1932), pp. 131 *et seq.*

catastrophic external event is likely to make it happen. In the degree that any society is global, it is an organization so tight that there are no loose elements whose rearrangement can be accomplished without sending tremors through the whole. No man in such a group can change any part of his institutional relations without affecting them all. He cannot make new friends or new enemies on the basis of new convictions without putting into jeopardy his whole range of affections. The tribe moves because the grass is shriveled and the herds are thirsty and the weather makes it necessary to change, or die. Or another fiercer or stronger tribe comes across the desert and absorbs the group into slavery which is partial conformity to a new pattern. And sometimes, but only rarely, an inventor puts a suggestion to the group which makes for quarrels, heartbreak, and progress.

It is astonishing to students who think of modern sophistications as "human nature" to observe how primitive groups can resist change even when it is clearly indicated and clearly beneficial. Part of the resistance of an Australian black to change is caused perhaps by his intense absorption in spiritual life. But it is still astonishing to see that the Murngin described by Warner [29] can actually eat the fruit of the mango trees that have grown in their hunting grounds from seed dropped by visiting Malays, generation after generation, and never once plant a seed to get another tree. Do they fail to see the sequence, seed in the earth–tree–fruit? Nowhere in the armor of their self-respect is there a chink into which the new idea can be inserted, and it would be interesting and valuable if we could know how many times an innovator has suggested planting mango trees only to be derisively howled down. Starvation, never far off, makes the savage relish worms but cannot make him plant for a distant harvest. Those visiting Malays, however, had learned a little agriculture at some simple stage of development. No society can be thought of as impenetrable. But the global kind is the least easily changed and when change does come it comes as a shock.

[29] Lloyd Warner, *A Black Civilization* (New York, Harper, 1937).

The mosaic type of society, if one could be found in pure form, would be a much more developed and sophisticated complex of institutions but it would be in some degree a combination of global groupings whose relations with each other would be polite indifference. This may be illustrated by the revelation made by an Austrian scientist of my acquaintance who even after six or seven years of life in America was still disturbed by pluralistic ways. He was once caught in an argument between another colleague of mine and myself and took my colleague's side. Afterward he apologized to me for differing. He would like, he said, to be "friends" with both of us. I pointed out that his disagreement with me on a matter of scientific opinion could not possibly affect friendship; I was on excellent terms with the other man who had begun the dispute, and would continue to be. But my Austrian was not comfortable. "When I lived in Vienna," he said, "I was a member of the X group. They were my friends; not men of other groups. I visited with them and took my recreation with them *because* we agreed." And this, I told him, helped to explain how difficult it was to found a free society in Austria.

A sharp contrast came to my mind as I remembered visiting once, years ago, at the home of a distinguished physiologist in England. I asked his wife, a sweetly dogmatic Tory, if she would invite over for dinner a scholar with whom I wanted to talk about education. This scholar was and still is a notable liberal. My hostess called him on the telephone with the most evident sincere affection and asked him to come over. But she added to me, "We'll have to dine early because he is going to some meeting or other, and God knows what treason he'll be talking." This was *not* the sign of a mosaic society but was pluralistic in my precise meaning.

The mosaic society, as in Austria, is the one in which different patterns of loyalties are tolerated but where those of similar patterns in any part strive to bring themselves into institutional contacts with others whose whole patterns are as nearly as possible like their own. The social grouping is mosaic in the sense that it is made up of smaller global groups over and above which there is some civic loyalty, and indifferent politeness, but for whose members there is little real contact with variety.

All societies are mosaic, of course, to the degree that they are divided into self-conscious social or economic classes. The Marxian dialectic is the product of a mosaic England in the early stages of industrialism and the hypostasizing of these groups into social "forces." From our point of view, it is always necessary to remember, against the self-deceptions of verbal habits, that any entity may be unitary or one part of a group, depending on our own subjective approach to it.[30] So any actual society under observation may be thought as more or less made up of individuals, groups, and combinations of groups. But we are classifying societies by the kinds of loyalty patterns that may be observed in the conduct of their members.

In a global social organization the patterns of all comparable persons tend to be similar, and if one member of the society feels an impulse to change, that will affect his whole set of associations, because he shares all his habits with all his friends and they share all their loyalties with all of his.

In a mosaic society there are differing patterns in the society as a whole, but men tend to organize their institutional relations into similar or identical patterns with their associates and tend to live intimately only with those who are similar. Here an impulse to change will cause less of a convulsion, because there is literally another place for a man to go when he wants to change, there are other groups within the society into which he can move and with whom he can find emotional accord. He may break with his family or his party or his church or his lodge but he is not therefore outcast, nor institutionally homeless.

In a pluralistic society there is still greater leeway for individual action because each individual is related to large numbers

[30] George A. Lundberg, *Foundations of Sociology* (New York, Macmillan, 1939), pp. 68-69.

of differing groups, his attachment to each group being a shared pattern of behavior in respect of one interest alone. He can have one set of friends in church, another set of friends in his work, another set of friends in his lodge, another in his neighborhood, another in his recreation, another in his party, and so on. Where there is leeway for individual action, there is a chance for inexpensive change. The pluralistic society makes freedom possible. It provides a range of normal choice and change can be rational, purposeful, and easy.

CULTURE AND SOCIAL ENGINEERING

I

WE HAVE TOLD WHAT WE MEAN BY A VALUE, FREEDOM, A GOOD TO be striven for without any proof that it is a good beyond man's determination to have it. A postulate. An absolute, if you like, but that is a dangerous and misleading term. We have told what we mean by scientific thinking and given reasons for trying to think that way about human action.

We can begin thinking about human action with consideration of the fact that such action takes place in a social context, in a "culture."

In a preliminary fashion we can think of culture as the sum of human achievements that endure, such as symphonies and ways of frying potatoes, philosophies and sports. Contact and initiation and education, as the organized institutions for that purpose, are what keep culture alive. As culture becomes more complicated, more of it is learned in school and in adult life. As it changes faster, more is determined by overt rules, or laws and less by custom. It is plain enough that men learn many things from their elders without conscious purpose, as we learn the beginnings of our "own" language, but we learn some of them intentionally, whether or not we make use of any institution. The special importance of education in social engineering lies in the fact that it is the special institution set up for the purpose, and whenever we set out to modify the processes, wherever and however active, that accomplish these teaching-learning results, we say we are changing "education."

If culture is what is learned, it cannot be material. But it has necessary material accompaniments that are the exterior inheritance. Among anthropologists and sociologists, there is considerable variation in the use of the word. Some have preferred to have culture mean mental facts only. Marett, and others as eminent, would prefer to call this "moral culture," the material tools with which it works, "material culture." The mental and physical facts limit each other and both must be considered. Here we shall call the mental facts of culture "institutions" and the material concomitants "tools." The important part for us to realize, whatever name system is in use, is that the continuity of any set of social achievements is dependent on the mental conditions, not the material. Men can continue to know how to sail if they have no ships. They are likely to forget, of course, the tools they do not use. But the art of navigation, a typical trait of culture, a typical institution as a set of mental facts, could conceivably be kept in continuous life by the teaching and learning of successive generations even if there were no ships or oceans, whereas the existence of all the ships that could be, would not maintain the art of sailing if no one had learned what to do with them.

This instance is highly artificial, intentionally so. Navigation is best continued when there are both ships and memories. But the artificial illustration makes it easier to see the truth in an actual instance. It is often said that the people who live in Pennsylvania have industries because they have coal and iron This is true in the sense that they would have to have the tools, including the machinery, to use whatever knowledge they might have acquired. But the Pennsylvania Indians lived for centuries on top of the same coal beds, the same iron ore and petroleum deposits, as are now exploited by the industrialized Americans They had the raw materials, the stuff out of which "material culture" could be made. What they lacked was the intellectual skills on which industry depends. The institutional side of culture is mental and its continuity is a mental process.[1]

[1] Albert Blumenthal, *The Best Definition of Culture* (Marietta, Ohio, Marietta College Press, 1937).

The importance of this lies in the fact that the continuity of mental processes is maintained by communications among or between persons and very largely by means of symbols. In a sense that is highly significant for social engineering, culture is a system of symbols and the behaviors which those symbols indicate.

The relation between institutions and tools, as the mental and material aspects of culture, may be put in this way: the physical environment sets the limits of cultural development. Environmental problems merely make possible, they do not provide, their own solutions. As Wissler puts it, "The influence of the environment thus appears as a passive limiting agency rather than a causal factor in the development of tribal life." [2]

It is important to oppose the easy determinisms of some exponents of "geopolitics" who would fix the destiny of nations by the historical accident of their home. Like all other rational, but unscientific determinisms, these need to be resisted on behalf of freedom. They are not sound interpretations of the "laws of nature" as they pretend to be. They do not merely describe coercions that are in the nature of things; they give excuse to merely human coercions that invoke them for support. A great deal of so-called "economic determinism" is subject to the same warning and the same rejection.

Both institutions and tools accumulate. They grow in number and complexity. The rate of change is itself changed and tends to quicken. Paradoxically perhaps, the more inventions there are, the more opportunity for inventors. The great leaps in human thought were those made in the primitive stages and, of course, this is still true in all those aspects of culture that are still primitive. But in developed phases there will be an acceleration in change. Inventions, though small in scope, will be numerous and incessant. At last, men in a later phase of culture, like our

[2] Clark Wissler, *Introduction to Social Anthropology* (New York, Holt, 1929), p. 339. Cf. A. A. Goldenweiser, *Early Civilization* (New York, Knopf, 1922); R. H. Lowie, "Subsistence," in *General Anthropology*, ed. Boas (Boston, Heath, 1938), p. 320.

present industrialism, may organize the invention of things and material processes as an institution in itself and set aside a portion of the social energy to keep it going.

There is no sign, however, that any society is ready to set aside in like manner a portion of its social energy for invention in social behavior. Industry is a thoroughly rationalized activity and the conservative rigidity that bound its ancestral phase, primitive agriculture, is long since gone. And industry is therefore inventive in some of the human aspects of its own business; timidly in relations between management and labor, more boldly in the relations between buyers and sellers. But not even in a highly industrialized society like ours is anybody professionally engaged in inventing new relations between husbands and wives, or parents and children, or even among friends. In these institutional relations there are strong impediments to creative power.

At the same time, a strong drift in the development of an industrial culture makes it possible constantly to enlarge the conditions of freedom. Mastery over material forces may not bring peace or happiness, in fact it may not even bring prosperity, because complex economic structures are so unstable; but by getting away from the stability of primitive life, we are also getting away from immediate necessity.

Man must eat, no matter how civilized, but simple life is both precarious and bound. In very early stages, most of the energy of any group goes into the food hunt. Sometimes a savage culture will appear to be content with meager comfort but will have enough energy for elaborate religious and ceremonial life, as do some of the Australians. But the food hunt must still take up nearly all the time and energy of tribes that live, as savage tribes generally do, on the edge of starving. And although they may not be hungry, the workers in any culture who do not have access to the products of industrial factories must lead a simple life, since the gadgets and comforts and services of a high standard of living are made in mass production centers. Farmers and miners live simpler lives than the factory workers in great towns.

As industrialism grows, less and less of social energy goes to the food hunt and more and more into those activities which, as "standard of living," enrich and complicate the life of the average man. The process comes close to current experience. In the United States, for example, we had nearly half of our population devoted to agriculture as recently as two generations ago, in 1870. By 1930, this proportion had been reduced to 25 per cent. In 1940, it was much less. And yet, in 1930, a larger proportion of our total population was engaged in some form of gainful work than in 1870, so those who had left farming had not left work. And the total population had more than tripled, so a much more productive farming system was giving us food and fibres and exports with the use of smaller human resources. We are not concerned to trace the explanation of this in the changes made by machines on the farms. Our point here is that the large numbers of men and women who had left the farms were making those goods and producing those services that made up the standard of living. They were building, selling, and taking care of automobiles and radios, elevators, beauty shops, picture magazines and motion pictures, packaged foods and kitchen machinery, and all the other things that their own ancestors in 1870 simply did not have.

II

We can see something of the nature of culture by noting facts of this sort about our own. But the concept remains difficult and there is much loose interchange of terms that deepens confusion. What, for example is a "society"? The term society is a noun of general parlance, meaning any organized group in which one set of institutions, such as language and government, prevails. It is an older term than culture in sociological writing, and nearly all the generalizations that have been made about societies can be applied, in so far as they are true at all, to what we now call cultures. It is important, however, to get rid of some of the hypostasizing that has been going on in thinking about society, the kind of thinking that makes society a thing in spatial existence, a so-called organism.

To call a society, or a culture, an "organism" may be useful as a metaphor. It is also dangerous. Whenever we make the painful effort to think scientifically about ourselves and our actions, denying ourselves the privilege of judgments of value, we can well go further and be skeptical of needless entities. We follow the logic of nominalism. Here, as in all such tricky problems of method, the adjective is less dangerous than the noun. Society is "organic" in the sense that the patterns of behavior that constitute the relations among people making up a society are interacting and interdependent. There is among them a "division of labor," so to speak, and this is related to specialized social functions of persons with which it can easily be confused. The specialized functions of persons are for the most part occupations; men have jobs. And in the interactions of their working they maintain each other as each maintains himself. But the relations in the social group that are like the organs of a living creature are the institutions that are in themselves relations among persons, some having to do with the feeding of the group, some with its political peace, and some with enlightenment. Institutions are "like" the organs of a living creature in having function and power. But it is the food complex, the food institution, not the farmer or the grocer, which is comparable to an organic alimentary system in a living thing.

When, on the other hand, we say that society *is* an organism we are creating something in imagination that has existence nowhere else. This is an hypostasis, harmless enough in common talk, or even in some phases of sociology. But it is ruinous to clear thinking when we are trying to build a structure of basic concepts. We are led into extravagances of myth making, we talk about the "group mind." We make heroic characters out of "societies" or "civilizations," as Spengler did, and make dramas out of history. Perhaps that is how history should be written; the historians will decide. But the sociologist will consider that Hegel and Spengler (and, as far as this goes, Marx also) were writing mixtures of poetry and philosophy. There is little science in that brew. Men had poetry and philosophy long before they did any scientific thinking, and it is quite possible that they will continue to create stirring and satisfying myths long after all science has been forgotten. Indeed, it is possible to contemplate, without tragic horror, the possibility that science is only a brief and sterile interlude in human annals, brief because objective thinking is too difficult, and sterile by token of leaving no trace. This is no reason for failing in the effort to use science while we are at it; the "organic" myth does not help us.

On the other hand, society is not a "machine," and we are not urging any "mechanistic" as opposed to an organismic view. A society is a set of relationships, not a "thing" that can properly be given such a name.

All these quarrels over names have an importance, however, because we allow our preferences in vocabulary to determine our way of work. The organic metaphor is dangerous because it leads to moral judgments and impedes science. For example, organisms, that is, configurations of matter organized in what we call "life," go through cycles of growth and decay. If in looking at a group of collaborating human beings, the observer tries to see, by the organic metaphor, a society that can grow and decay, he will impose morphological sequences for which he has no evidence in the social facts. In forcing the observable facts into conceptual patterns he will be adding an imagined element, the "age" of the society. Hence, we will get such notions as Lewis Henry Morgan's series of inevitable phases in social change, or good Marxian determinism. We shall then speak of a primitive culture as a "young" culture, although there is no certain reason for thinking that any one group of persons now living inherits a shorter history than any other. We shall then, in politics, demand the sacrifice of living and suffering human beings to myths of state. Adjectives applied to society, or to culture, are no longer harmless when they lead us into crime.

Human action must take place somewhere, sometime, and always as the exemplification of some phase of culture. For our purposes, we need more searching descriptions of culture and more complete hypotheses as to its nature and its dynamics. We are discussing the management of social action. Management would be a series of engineering interventions in culture for the purpose of making changes, hastening some changes already under way, checking others in so far as might be possible. The use of science in the management of human affairs would have two phases, not always separable. They are exact description, or economic summations of description in natural laws, and a second phase, the manipulation of events in order to bring into reality a chosen one among many possible and predictable events. And although science is a typical product of cultural activity, embodied in institutions and in personal habits, the whole of cultural activity may be called "engineering."

By engineering I mean a conscious intervention among the processes of nature, taking advantage of some processes in order to check others. A good example would be checking the natural process of the spoiling of fruit by an intervention that chemically produces cold. It is as natural for a peach in a refrigerator to stay fresh for a while as it is for it to rot on the ground in summer sun, but it took a conscious, that is a purposeful intervention to make the difference. In this sense all cultural activity is engineering.

Nature is, to our human view, wasteful of individuals, careful of processes or continuing patterns. The impulse to produce one's kind, for example, is tremendously overloaded. There is a marginal amount of wreckage among human beings in culture caused, for example, by the sex impulse. No moralizing, certainly, and possibly no intelligence can prevent it.

Men, in culture, unlike nature as a whole, can choose to be careful of individuals as well as of processes, or even more careful of individuals, making, as in democracy, the individual the end to which the process is directed. This is diverting the trend, of course; and that is what culture sets out to do—to divert the trend.

If it were necessary to have a naturalistic theory of freedom, to found the love of freedom in some general process that we can see implicit in natural change as we can read the history of natural change up to now, we could find it, I think, in this way of looking at culture, as engineering. The concept of culture set forth here leads to the theory that freedom is the natural purpose of cultural behavior. The theory is worth study because we may find it necessary, in some future state of advanced scientific rationalism, to have objective proofs of the worth of our values. No such proof is necessary to our argument here because we have used freedom as a postulated good in order to illustrate a method. This implies that the method, properly used, might get for us in other circumstances some other desirable condition. The theory is stated briefly not because it is necessary but because it has in it a kind of reassurance that philosophy need not disdain.

Culture is engineering, or intervention in nature, for the ultimate purpose of saving individuals as against processes, and that is essentially the purpose of freedom.

In a recent attempt at a scientific theory of culture,[3] Malinowski calls organization, meaning the organization of human beings in groups, "the essential fact." This statement is doubtless true as it stands. And Malinowski was a great pioneer in the struggle to dig social science out of literature and make it science in a real sense.[4]

But in spite of Malinowski's genius and his labors, there is already, almost before his posthumous books could be gotten into print, an old-fashioned sound in his vocabulary. He speaks

[3] B. Malinowski, *A Scientific Theory of Culture and Other Essays* (Chapel Hill, N. C., Univ. of North Carolina Press, 1944).

[4] Lyman Bryson, "Writers: Enemies of Social Science," *Sat. Rev. of Lit.*, XXVIII, No. 41 (Oct. 13, 1945).

of form and function as the terms in which the ideal isolates of culture are to be described. The material of biology is something like the material of sociology; they both deal with living creatures some of which live in organizations. This does not, of course, compel us to use biological terms. But form and function, as used by Malinowski, are biological terms and there should be congruity in the terms used by biology and sociology on the highest levels of generality. So, form and function sound somewhat old-fashioned as a conceptual framework for describing social facts because biologists now speak rather of energy and organization. The extent to which the entities have remained the same conceptually, while the vocabulary has changed, does not make much difference. Energy and organization are better terms for the description of culture than form and function.

Thus, borrowing conceptually from Malinowski but changing terms, we can say that culture is human energy organized in patterns of repetitive behavior. If the patterns are not involved in predictable uniformities they do not exist as observable phenomena, and Malinowski is right in saying that cultural facts must be publicly observed. Patterns in human action can be seen in the same way that a melody can be heard. The mind puts successive experiences together into a single experience that has a rhythmic shape. We can see the organization in patterned human actions; we infer the existence of the energy that is, as we say poetically, "expressed" or expanded in those actions. More strictly speaking, the physicist refuses to define energy because he says "no general meaning can be given to the energy concept, but only specific meaning in special cases." [5] But we can think of energy in a less mathematical sense and in the special cases of observed human movement. In this way we have ideas sufficient for our purpose. And we can discuss culture as a pattern of social engineering activities which

[5] P. W. Bridgman, *The Logic of Modern Physics* (New York, Macmillan, 1932), p. 128.

have as their purpose the realization of liberty in the actual lives of persons.

The basic difference between democracy and other patterns of human organization may be put, generally, as the difference between a system that sacrifices institutions and other processes in behalf of the human life that is actually endured by human beings, as opposed to the systems that sacrifice persons to reified institutions. The first, sacrificing or using all processes to save the individual, is democracy. Sacrificing persons to processes is exemplified in our time by the "fascism" that kills the personal spirit for the sake of the state. Democracy is a more developed form of culture than any autocratic system could be, in spite of the fact that autocracies have developed often, both early and late in culture history.

Culture, as social engineering, intervenes in natural processes for the purpose of saving the "particular," which nature wastes, without destroying the "kind," for which nature is willing to disregard the fate of particulars except as vehicles of kinds. This is another way of saying that culture is always development toward greater development of individual persons. But institutions, as collective action, are the means by which culture works to its ends. Institutions, as we all can see, tend to harden and lose their effectiveness for the purpose of developing persons. In their rigid form they are usually instruments by which some individuals interfere with the liberties and growth of others, in the name of the collectivity. This is corruption. Democracy is the form of social order least likely to permit these rigidities and corruptions to get fixed in a society. Hence it is a system in which culture can most effectively continue to serve its basic purpose, saving individuals without destroying kinds.

Hegel's definition of liberty, to live according to the laws of one's nation, is thus only a sophistication of freedom. One must know the nature of those laws to be sure that obeying them is liberty. A society that was in full enjoyment of liberty as Hegel defined it might be a good society. That is, it might be successful by the calculus we are considering, and it is obvious enough that for human beings who hated true freedom it might be a better society than a true democracy. It is in these fringes of creative power with which the human personality is endowed that we make the sportive choice. We disagree with Hegel. What we are, as the result of our capacities and our histories, makes us, Westerners of the twentieth century, choose the kinds of laws that allow us to create, not merely to conform. And with those predilections (a thinker like Mannheim would say because of them) we believe that we see a process toward freedom in the development of culture when it is thought of as engineering, a purposeful intervention among the processes of nature.

This problem, which may turn out to have great philosophic significance, is difficult to discuss in the context of modern times because technological change and the drift of ideation both have the appearance of a movement toward collectivism and we cannot now discern what effect collective ways will have on individuality. The only creature that could agree with an ant that perfection in social organization was to be achieved in perfectly regular collective action would be another completely social insect, a termite or an ideal communist. Flesh and blood communists and all who question or resist the collective trend want to know what forms of freedom the trend allows. They are suspicious of reversion to those easy ways of solving social problems that were discovered first in human history; the unity that was mere poverty of desire, and the emotion of belonging that was not much more than a deep quiver at the smell of the tribe.

We believe that the development of culture has been a development toward higher forms of individuality. In the beginning, persons could not differ. It was not safe then to allow differences; global loyalty was the only kind that could survive.[6] In order to intervene successfully in the net of processes by which he lived

[6] P. W. Bridgman, *The Intelligent Individual and Society* (New York, Macmillan, 1938).

in the world of nonhuman nature he had to intervene along the lines of less resistance. His engineering techniques were directed toward the achieving of the only kind of persons then possible, simple and undifferentiated primitives. He worked with natural laws by harmonizing his soul and body with the world he thought he understood. The strengthening of his techniques, the growth of his engineering powers, has led to a constantly more masterful intervention, which is another way of saying the building of constantly more complex and richly diverse cultures.

This diversity has made possible the ultimate achievement of richly diverse persons, true individuals, differing not only from those in other branches of the common civilization on the earth, but harmoniously diverse as variations in single groups so that the world of human differences is not only a mosaic but a web. In religion, the highest forms contribute most to this basic purpose of culture; they enhance in each individual his conviction of the cosmic significance of his own soul. They reassure him against the terrifying world of nature. They helped him to be at home with the spirits of the wood in his primitive stages; nowadays they give dignity in the face of the vast impersonality of galactic space. The advance of any basic phase of culture is the increase of individuation, and democracy is, philosophically understood, a refinement and an explicit form of that purpose by which we move more swiftly and surely, because consciously and inventively, toward the orchestration of lives.

III

There is great danger now that the wrong step will be taken in the development of social thinking into social science. Men who can make a proper diagnosis can offer inappropriate remedies, as does Robert Lynd.[7] Professor Lynd has had some interested support from educators (not so much from sociologists, it seems) for his assertion that in learning more exactly defined

[7] Robert S. Lynd, *Knowledge for What?* (New York, Harcourt, Brace, 1929).

and measured facts about the behavior of men in society we should know, while we are learning, what we are going to do with the knowledge gained.

The proper answer to this error was given long ago by Herbert Spencer when he pointed out[8] that knowledge can be dependable only when searched out for its own sake, but we need for its application in society an art of "social engineering." It has been well stated by Thomas and Znaniecki that

> there is no doubt that actual situations must be handled immediately, we see that they cannot be solved adequately as long as theoretic reflection has their immediate solution in view. . . . And such a science can be constituted only if we treat it as an end in itself, not as a means to something else, and if we give it time to develop along all the lines of investigation possible, even if we do not see what may be the eventual applications of one or another of its results.[9]

The fact that anyone can make, as Lynd has done, a scientifically naïve statement, and be praised for it, indicates how far social thinking is from science. In mature sciences, those which have outgrown magic and wishfulness, a clear distinction is made between the problems of theory and the problems of use. There is, to be sure, a fruitful interchange between the two phases. They are both "scientific" in any real meaning of that word. But they are not the same. The physiologist and laboratory experimenter, for example, stick to their last, which is to add any fact, no matter how minute, no matter how "insignificant" it may seem at the time, to the common store. Syntheses are made, of course, and men are often seeking syntheses. They work from hypothetic guesses as to what the facts may turn out to be. They work with ideal types toward "laws." And, even in medical research, it is often said by competent investigators

[8] Herbert Spencer, *The Study of Sociology* (New York, Appleton, 1929), Chap. I.

[9] W. I. Thomas and Florian Znaniecki, *The Polish Peasant in Europe and America* (New York, Knopf, 1927), I, 15. Reprinted by permission of the publishers.

that the direct attempt to solve the theoretical aspects of practical problems is the one attempt most likely to fail.

On the other hand, the practicing physician, who is an engineer in our terms, goes at the solution of a practical problem as directly as he knows how. It is a useful although partial statement of the difference to say that the investigator uses a method and accepts the results, whereas the engineer uses all possible methods until he has achieved a result that was predetermined.

And yet, all science is for use. The difficulty, to repeat, is not in the diagnosis but in the remedy. The trouble with the social sciences is not, as Lynd and others have said, that investigators are not thinking enough about the significance or use of what they seek. In education, at least, I can say from my own long association with doctoral investigators, they are far too much concerned with applications. The trouble lies in the fact that in social thinking we have so far failed to make the necessary distinction between theory and engineering that we have almost no orderly training anywhere that can be gotten by the would-be engineer. Teachers might be, in this sense, engineers who use the social sciences and psychology derived freely in laboratory and field work, for the achievement of clearly seen practical ends. But teachers are as yet confused as to what the term "education as science" might mean. Social workers are another possible group of engineers, but they do not seem to see a clear relation between their own work and theoretical knowledge of the pathologies of social groups. Nor do politicians learn their business from political science.

The great need, the crying lack, is clear definition of this work of application. If there were a clearly seen function of social engineering, the function of thinking about culture might be attended to with clear conscience and some hope of scientific success in the study of human action. I cannot say what this would do for the morals of the social scientists who would look to the problems of theory, although scientific investigators appear to be rather high-minded men in general. It might not give them as great a conviction of virtue as they now have. But we can be sure that the human benefits derived from their work would enormously increase.

We look to science as an instrument for good. But science as a method, or predominant social habit, is on trial before mankind. This is true far more than most Westerners, satisfied in their cultural parochialism, can realize. There have been flashes of the scientific spirit in other cultures, but science in any real development is a trait of the West, of Europe and the United States. The gadgets go over boundaries, of course, and advanced communication and transportation, airplanes and air messages, may spread actual use of machinery into remote spots. But science is a spirit, a set of mental habits, an institution for investigation and control, and it is, or has been up to now, a Western character.[10] Since they have never shared the habit, the rest of mankind, in fact the most of mankind, may be simply not convinced.

Even in Western cultures, we know, only a small proportion of the people have any more of science than a knowledge of which buttons to push. And those who can think objectively about material things are seldom able to advance very far toward a scientific control of their own or their neighbor's behavior. However, all of us in Western cultures are controlled to a large extent by the scientific thinking of our cultural companions, by the dominant ideas of our groups. Not so for China and India and other scattered millions who may never accept science at all.[11]

If human approval is finally won by the institutions of science it will certainly not be by material power alone. Only the kind of scientific solution of human problems we are here discussing can get and hold the world's belief. In fact, the objective manner of thinking, which is one pattern in the mind's normal functioning, is always on the defensive even in the most favorable cultural situations. It will not only fail to dominate, but will probably

[10] Cf. E. Zilsel, "The Sociological Roots of Science," *Am. J. Soc.*, XLVII, No. 4 (Jan., 1942), p. 544.

[11] Lin Yutang, *Between Tears and Laughter* (New York, Day, 1943).

be defeated and discredited unless it does better than it ever has done up to now in saving men from tragedy and despair.

To those who believe in the ultimate helpfulness of the mind's power of objective organization, this defeat would postpone the triumph of the human spirit over secular problems, if indeed it did not make that victory forever impossible. If we are to persuade men to make this effort, we have to meet not only the pessimistic fact that objective thinking has done only a little thus far to make human behavior better by any rational standard, but also to combat the widespread belief that science is materialistic, deterministic, and inhumane. It is materialistic, they say, because it has dealt successfully only with material things. It is deterministic because it tries always to establish the repetitiousness in things. And it is inhumane because it disregards values. We shall believe in a scientific humanism when we believe that values can be achieved in fullest measure if objective organization of experience is used to understand the repetitiousness in human actions (which is there whether we like it or not), that this can best be done by the methods of material investigation, and that the creative powers which invent and fulfill values will thereby be released.

THE ENTITIES OF SOCIAL SCIENCE

I

IT WILL BE USEFUL NOW TO RETURN AGAIN TO CONSIDERATION OF our purpose. These are prolegomena, notes to serve as introduction to a social philosophy, scientific humanism. They are not social science but philosophic argument about it; not social engineering but an attempt to prepare the way for it. For this purpose it was of great importance to state the theory of culture as engineering in behalf of individuality. But since we postulated freedom, the state of individual growth, as the good toward which our engineering efforts would be directed, it is important now to speak not of culture as engineering but of the engineering of culture, to discuss in nontechnical terms what the social engineer may be expected to do. We have asserted that science and applied science can be used to create the conditions of freedom and have tried to establish some general ideas about the nature of social change that will give a context in which the rational manipulations of social change can be considered.

What would the social scientist actually do? It is clear that he would continue to do a good deal of what has always been done by all the kinds of thinkers that have been trying to understand humanity, the poets and moralists who contribute to what might be called the "natural history" of mankind. The natural history of a subject is the description of its subject matter in its natural state. This kind of description needs technical terms for classification but deals with no artificial or ideal entities and has no special concepts. Let me illustrate from the field of biology. The natural history phase of the study of *mus musculus* would be the observation of his habits and his kinds. It would require the stalking of the living creature; the units of observation, the entities described, would be the organic unities of the mouse's world. Out of such study we would know the normal life cycle, the habitat and the organic pattern of each species but we would never learn anything about any scientific problem of the theoretical order, as for example the mechanism of mouse inheritance.

Morgan probably knew a great deal about the natural history of the fruit fly. He knew the kinds and the habits, and he knew tags for the kinds, like *Drosophila melanogaster*. At least we can presume that he had such knowledge but it really makes very little difference, because if he had had all the knowledge in the world of that sort about fruit flies, it would not have helped him to understand heredity, which was his theoretical problem. For that he had to work with some such concept as the gene. The geneticists had to create (or discover) an ideal entity, on a level of abstraction that made precision possible. It should be noted again that the gene may not, in one sense, even exist. It is an ideal entity that may be manipulated symbolically and it is of such a nature that the results of such manipulation, if logical, may be paralleled in observable facts. In other words, "laws" can be made and also predictions, and observations will "verify" them. For scientific purposes it is not necessary to go further with any judgment as to existence, reality, or truth.

No one doubts the importance of the natural history of human beings in society. That accumulation of knowledge has been added to, intentionally, by almost everyone who has ever written or spoken seriously in any form of literature. And it can be useful in social science in the same way as it is in biology. The difficulty is in taking the next step. As long as biology was merely natural history, the mastery of biological events was slight. The mastery of biological events is, of course, the deliberate production of varieties of animal life. Thus the proof of theory is in engineering. Now, to take these steps in social knowledge, to go from natural history to theoretical ideal description and thence to the production of new varieties of social life, is not an analogy but an exact parallel.

Resistance to this extension of science in dealing with human behavior takes the form of skepticism. In cannot be done, we are told. Human beings are different from animals and plants. Or, they are like the other living things but too complex for such study. The skepticism is justified, of course, as all caution is morally justified, but the answer to the resistance is to make the trial.

II

It may be useful to explain, while we discuss the definition of entities, why it is that writers of all kinds of literature are the enemies of social science. Most poets are in awe of physics, or of their own image of the physicist's ideas. But many writers whose primary aim is the expression of thought and emotion in artistic forms, poets, novelists, and critics, would prevent the study of human behavior from taking on the same qualities as have made science out of the study of other natural phenomena. Much of the resistance is instinctive. Few of the writers who will make no effort whatever to comprehend social science know their own motives, although these motives are good examples of the subtle springs of human action that artists take such pride in understanding.

Why are writers the active enemies of social science? When one looks at the slow transformation of cultural forms the enmity is not surprising. It is natural. As a culture develops in complexity and richness, it develops also in the division of labor and social science would have to take away from the writers part of what the writers think is their job. The attainment of scientific thinking about ourselves, as persons, is one of the last stages in the development of the division of labor. It will come in spite of the fact that the writers resent encroachment unconsciously and resent with explicit scorn the use of any other methods than their own in dealing with what they think are their own peculiar questions. When it does come, this division of labor, writers in the forms of literature will not have lost anything but their claims to know the only valid kinds of description of human behavior.

We can attempt to state what the division of labor might be. The distinction between artist and scientist is difficult. We have been trying to describe scientists. The scientist, right at the beginning of his operation, disappoints and alarms the artist because both deal in words and their attitude towards words must be quite different. The word "artist" is here used to designate one who conveys emotion by means of design. This is the essence of fine art, to convey emotion by means of design. Obviously, most artists do a great deal more than this, conveying emotion also by direct representation and by immediate appeal to the senses, and they convey by all their inventions much more than emotion—extra valuable baggage in ideas, judgments, and information. But his share of the labor as artist is to convey emotion.

Literature, of course, is more than the writings of the practitioners of a fine art. It always has been, probably always will be, a much wider category of human communication. It is a vast matrix, the undifferentiated mass of human expression by one kind of symbol system. Out of it as the division of labor has developed have been carved philosophy, much of religion, much of the "physical science," even. Homerian epics were partly divided into religion and philosophy and science by Plato. The modern social scientist is only a late comer to the ancient hunting ground. But he brings his own weapons and hunts new game.

The literary artist, in so far as he is at the moment practicing his business, uses words for their suggestiveness, their color, the wide possibilities of partial understanding. It is absurd to suppose that a communication between an artist and one of his appreciators is ever exact or ever limited to that small area of real experience they may have in common. To the appreciator, the artist's words are symbols of the appreciator's own experience and they evoke out of him only what his own experience and nature provide. This is not to say that the artist does not have a meaning; on the contrary, he has many meanings, overlaid and intertwined patterns of meanings, and his words are signals from his private world to the private world of those who enjoy his work. An artist is at least many things to many men.

As a matter of fact, although this again is a digression, it is interesting to note that the greatest artists in words are often notably deficient in the power of abstract thought. This is not what makes them interesting but it surely does not detract from our appreciation of their writings since it is involved in their complex emotional and sensitive temperaments. Tolstoy is an excellent example and we can cite him without making invidious remarks about living novelists.

The scientist by contrast has to deal with words not as the natural growths of man's symbolizing power, but as arbitrary symbols. So, naturally, much of the hostility of literary people for social science is expressed in criticism of what is called "jargon." Although writers are specialists in language, none of them will bother to think about the problems of terminology faced by the social scientist. In the development of any science, there comes a time when ordinary words will no longer do the work. The ordinary word has too many meanings. The scientist must have a symbol that will mean, in so far as this is possible, precisely the same thing to anyone who uses it in the scientific context. Such words are practically nonexistent in ordinary language. Ordinary language is made up of words of vague, complex, multiple, and oblique meanings. The literary artist is very seldom, if ever, interested in reducing the semantic load of his tools. As has been quite adequately shown by Richards,[1] he gets his effects far better by reason of multiple meanings. The social scientist cannot use, as the artist does, the language of every-day speech for the statement of a proposition of scientific importance.

This, I would like to say in passing, is not the same thing as "writing good English." No one could well defend the heavy

[1] I. A. Richards, *Practical Criticism, a Study of Literary Judgment* (New York, Harcourt, Brace, 1929).

and ugly writing in which so much of our scientific thinking is expressed, but when the critic complains about bad writing he is not getting at the root of the matter. It is still true that the scientist cannot put into the ordinary loose and "well understood" words a proposition that is to him of any importance.

Which shall he choose to do? Take ordinary words and give them arbitrary precision and singleness of meaning? Or invent new words? In either case, he is in trouble with the practitioners of literature. If he uses common words with special meanings, he is scolded for his presumption. If he uses newly invented words, the cry of jargon is a swelling chorus. It does him no good at all to point out that he is only following timidly in the footsteps of the scientist who deals in physical matter, in so-called physical facts.

The literary critic does not read papers in biology or physics because he does not know what they are talking about, and he makes no claim whatever that they ought to be intelligible to him when he will not take time to learn the vocabulary. But the social scientist is dealing with exactly the same material that the literary artist claims for himself, the behavior of human beings. The artist's own statements are, in his own opinion, "true." That they cannot be true in his sense and also precise in the scientific sense is a fact that he does not care to consider or finds distasteful when he does. He believes in what he calls insight, not in statistics. Insight is, no doubt, a way of perception of great value and of very ancient use. Insight can give us sympathy, understanding, and when used by an artist, can enlighten and inspire. Statistics are dull but more reliable. Jargon, or special vocabulary, can convey precisely the same meanings, for operational purposes, to a great many different minds. Scientists can get on with common words arbitrarily defined; for example, the biologist arbitrarily uses the word "organizer." Or they can learn precise signification for a set of made words like electron, neutron, gas.

It would be more than an act of charity, indeed it would

justify the literary critic's claim to special knowledge, if instead of scolding he would give some help to the scientist in his dilemma. The social scientists believe that we can have scientific description of human behavior and trustworthy predictions in the scientific sense only as we build adequate taxonomic systems for observable phenomena and symbolic systems for the manipulation of ideal and abstract entities. The problem might be solved by a curt notice to the literary critic that he should never try to read social science anyhow, any more than he tries to read the journals of genetics or molecular physics. But that, besides being impolite, would probably be useless, because literary people are quite properly interested in everything that deals with humanity. It would be far better, if he could, for him to understand how the scientist must do his work.

However, the general problem cannot be entirely solved by straightening out this matter of words. Those who claim insight will still be jealous of those who trust statistics and accumulations. This is a natural jealousy, not only because of encroachment but because we have lived so many centuries in which literature was the only wise and rational comment on human action. Aristotle knew what science was but he still had no sociological content in a scientific way. And for twenty centuries and more after Aristotle's time, all the wise, rational and serious thinking about men and their relations with each other, about societies and their destiny, was in literature. The whole lore and treasure by which our spirits are shaped are in those poems and plays and stories.

All that the social scientists can claim is that piety does not require us to go on forever depending wholly on an institution because it has sustained us in the past, and that this applies to literary psychology and literary sociology quite as well as to other institutions. Literary psychology (the term is Santayana's, not mine) can be practiced as long as there are competent writers interested in serious speculation about the workings of the mind. Scientific psychology is becoming something quite different. We can avoid the arguments as to whether Freud was scientist or poet, but we cannot get very far in a psychology that can stand the final test of trustworthy prediction if we call Racine a psychologist or expect Thorndike to be a poet.

Living men, of course, are mixed. It should not be necessary to say this but doubtless it is; all men are poets and all men are scientific but we are trying here to describe man *qua* scientist, not man *qua* man. We can well leave that to the philosophers. All we knew about mankind up to a few generations ago was in literature. What we know about mankind in the future will probably be in literature too. If important parts of this knowledge exist also in the form of scientific propositions, the division of labor will have been achieved.

There is one further obstacle, however, that needs to be talked about, and here the attitude of anyone seeking truth in judgments on human action is compelled to be severe with literary craftsmen. Writers themselves, and the critics and appreciators of writers, constantly make a mistake that is likely to be unfortunate for the appreciators of literature and fatal to any sound learning about human beings. The mistake is to confuse expressive power with dependable knowledge. There are easy examples of expressive power without much knowledge and for that matter without much content. Victor Hugo's talent will do for one. And there are, of course, examples of powerful expression weighted with powerful thought. In English, the work of Hobbes or Donne would serve as example. But, if one can say it without *lèse majesté*, there are many passages in the writings of great poets and even of great philosophers that are profoundly moving and that seem to be profoundly informing, whereas they are as false as they are beautiful. Does it make the point to say that they may be great literature in spite of being false but that by this test any scientific statement would be valueless? Expressive power, by any theory of aesthetics, is part of what makes an artist great. Only in the opinions of other artists and in his own mind does it make any artist infallible.

Going back to Tolstoy, we can safely call him one of the greatest artists that ever lived. But he wrote, in his tremendously expressive way, a number of volumes that are by standards of abstract thought mere nonsense. Evidently, a great novelist may or may not be a profound thinker. It is very hard to get artists to admit this distinction, since their expressive power is their chief gift and they want as much authority in the world as they can possibly get. I have no intention here of questioning the validity of what is called artistic truth. I mean only to say that artistic truth and scientific propositions are quite different things, serving different uses, coming out of different processes and attainable in general by different kinds of minds.

It would be unreasonable to expect that artists, being self-centered and lyrically gifted people, could accept quietly the scientist's invasion. There is little comfort perhaps, but there ought to be some in the fact that the scientist, as he becomes more and more scientific, writes necessarily more and more for a special audience. He cannot, like the artist, make his own audience. He is not working for appreciation but for operational validity.

If the poets could see how much the importance of their own work is enhanced by the services of the scientists, rather than diminished, they might be glad that certain tasks are taken from them. They can be thankful that certain tasks, in very specialized forms, are now in the hands of specialists or soon will be. The scientist can describe the behavior of human beings in precise terms and predict future events. The engineer can take this theoretical knowledge and make applications of it in control, since all science is ultimately for the purpose of mastery. But mastery itself is only the instrument of other purposes. Where do we get the aspirations and the goals to which a more rational society would undertake to go? Out of religions, no doubt, and out of philosophies, but also in great measure out of the world's literature, which includes so much of religious and philosophic thinking and so much besides of human longing, of human hope.

Scientific ideas are, of course, inventions. Ideas of value, ideas by which we live, by which life gets its meaning, are also inventions. The artist, as critic of life, is a critic of values and if he is great enough, he is creator, too. The truthfulness and honesty of his observations will be one of the things that give him authority, but his artistic authority comes from his power to express the revolt, the forward-striving, the development, and the triumph of the human spirit in making good the ideals that great human beings can conceive.

III

The question for us, now, is this, What kinds of ideal and abstract entities would the social scientist deal with? What would be the equivalent (not the analogy) of such an entity as a triangle or a gene? We can explore the context.

A social scientist would never hesitate to spend time and heavy labor on trying to be quite sure that he knew precisely what it was he was talking about. To get some sense of the kind of thinking he would have to do we can try to get to the exact limits of the entities with which he might want to deal. First of these is the unit, the one, the self, the human being. How can we change "human being," an imprecise and unmanageable term, crusted over with sentiments and values and suggestiveness, into something we can use? We want a working concept that will include all that can be useful to our own thinking and exclude all that is neither useful to our own thinking nor influential in the events with which we are concerned. This will not describe human beings as unique selves. They will be treated as abstract entities in scientific discourse because science, in its most advanced stages, deals with nothing else and can make generalizations only about the artificial or ideal entities about which generalizations can be true. I will call this first unit the "person."

Exact description of a "person," in spite of all our endless chatter about ourselves, will be strange. We have for so long thought of human events as different from all the other events humanly observed that simple factual statements about man are bizarre and repellent. It has been only a few generations since the public guardians of other men's consciences permitted us to return to earlier naturalism and think of man's animal nature as a normal segment of the animal nature of all the rest of creation. Even now, there is a widely and emotionally held belief that man's animal self links him somewhat disdainfully with other creatures and that he has essential spiritual differences from all other living things. On this the scientist, as scientist only, has nothing to say. When he thus is quiet, his silence is in itself shocking to some minds. Perhaps a statement of some of the anthropological ideas of the person will provide what is needed to make further discussion free of unintentional impiety. Its purpose, however, is to work toward a vigorous definition of an entity about which naturalistic statements of fact will be possible while uniformities in its way of acting are being observed. We have to find our own way of looking at persons that "have" bodies and "have" minds.

An anthropologist is a student of society who remembers that man "has" a body. All social facts, to his eye, have a physiological side. All human behavior is partly physical in the grosser sense; that is, we can behave only by doing something in the physical world which we inhabit, by means of physical bodies and on physical terms. All behavior may also be "merely" physical in the sense that there is nothing existent but the physical facts; this question does not here concern us. But the anthropologist, the student who has looked long and carefully at simple cultures or tried to reconstruct the beginnings of institutions, remembers always that men must eat and procreate, grow and work and rest and die—or they cannot be men at all. And he considers both their physical needs and their physical equipment when he tries to understand their ways of behaving. For our purposes, we need some usable general ideas about men and their behavior that will be conducive to thinking of ourselves in terms that will make social engineering possible. The anthropologist is, in some degree, a social engineer, and his descriptive methods can be of primary use to us.[2]

Man's great advantage over other animals is his meager range of instincts. He has practically none. Some psychologists have built houses of names, asserting that men do this or that because they have an instinct for this or that; whenever men do the opposite, that, it seems, is because they have also a convenient counterinstinct to do that opposite. *Let* instinct here mean what it does in biology: "forms of behavior which were not acquired by an animal through experience and learning, but provided for congenitally in the animal's structure, and manifested in the first occasion of performance, if not perfectly, at least adequately to secure the survival of the individual and the continuance of the species." [3]

Whatever else he may have that psychologists may wish to call by the same name, man is an animal almost entirely lacking in forms of behavior "provided for congenitally" in his structure. He must and does learn nearly everything he needs to know. He is almost endlessly modifiable; that is, he can learn almost anything. He is young a long time and must be protected until he knows enough to survive.

Man survives in our civilization not by merely getting strong enough to stand up and fight. He must also know which buttons to push and which words to use and when. Elaborate sets of fixed forms of behavior, instincts, not only do not need to be learned but cannot be discarded. If men were creatures so durably built, they could have only limited and unchanging culture, nothing like any culture now known to man or imaginable in the past. The chief difference between animal societies, the anthills for example, and human groups, is that the ants have not changed for thousands on thousands of years. All human societies do change.

[2] See especially E. D. Chapple and C. S. Coon, *Principles of Anthropology* (New York, Holt, 1942).

[3] Drever, in the *Encyclopedia Britannica.*

Man is really not so much the "rational" animal as the animal that never lives as his ancestors lived, and instead of saying "you can't change human nature"—if you mean human ways of behaving—it would be truer to say you can't keep human nature from changing all the time. Anything "human" is the result of a physical being, limited and endowed as it may be, living by some one of the possible cultural patterns. Cultures do change and men, because they are not fixed but highly modifiable, take on all the different and shifting patterns of their humanity.

What I am saying here has nothing to do with the question of individual stability. All that is asserted for the moment is that the physical creature who was given the opportunity to become a Cro-Magnon hunter 25,000 years ago, and did it with success, that same creature, the same in all his "physical" being, can today be a New York stockbroker in an equally natural way. The New Yorker of today and some very primitive man of many thousand years ago, Cro-Magnon, Borreby, or some other, are biologically the same. That is the point of importance because we are concerned with the fact that practically all, if not all, the complicated differences between them are cultural and are caused by their different life experience. In thousands of years of interbreeding there has been modification in the race, of course, some family strains have been bred out, others have been developed, and European white man is a racial bastard about whose parentage there can be much dispute. But each "normal" child born is as capable of savagery as of civilization. What makes the difference is where he was born, and when.

Man is physically as gifted as are most of the higher animals. His senses are on the whole as sharp as those of the other beasts. If he cannot hear as well as an elephant he can see much better. The dog enjoys both sounds and smells outside his master's range but the dog often cannot see what his master is pointing at down the road. Modern man may badly neglect his senses, especially the sense of smell which he is usually ashamed to enjoy, but his children continue to be born with all equipment for learning to hunt mastodons if any should appear.

Mentally, however, men are different from their brother beasts. The nature of those differences is still moot. Even if one leaves out the sentimentalists who reassure their own inadequate humanness by reading humanity into the honest bestiality of their pets, there are still many who see "thought" in the behavior of the simpler animal minds. We need not enter that tourney except to say that whatever the animal intelligences can do, they cannot construct a culture. That is probably because they cannot use symbols.

Whether the difference in mental power between apes, for example, and man is one of kind or only of degree does not matter; it is very great. A mechanical process, a bodily structure, only partly complete in the shambling ape comes to success in man. The brain is free. Instead of hanging at the end of a horizontal beam, the backbone of a quadruped, the head is stuck on the top of an almost upright pole. The four-footed creature has a brain bound in by the muscle and sinew that wag his head and hold it on the end of his horizontal spine. The upright human being has no such stricture on his brain box. Our feet, marvelously balanced machines, enable us to stand squarely upright and the posture is not so solid as is the four-footed, four-point landing of the horse or cat and not so good for high speed, but it makes possible a looser attachment of the skull to the body and that looser attachment makes a freer brain. The hand can tear food into bites and that relieves the jaws which then need no heavy ridges over the brow to support their hammering and that also frees the brain.

Even one who is not specially interested in the anatomical changes in evolution can well ponder these details in the skeletons of apes and men. They cannot be understood without full explanation and actual handling of skull bones. But they show, when their reason is fully grasped, that man is the end product of changes that sacrificed other constructive advantages to free the brain. And so the human brain is a greater portion of the body weight than is the brain of animals less specialized for learning.

The great capacity for learning is a function of that larger node in the human nervous system, the cortex. All learning is in some way, still mysterious, a modification of living matter, and we do not need to define strictly either life or matter to operate with that idea. Man can be seen to learn more, much more, than any of his companions.

As we step from the physical realm into what we call the mental, we get at once into trouble. We can avoid one philosophical puzzle by asserting at the start another definition: *Let* mental stand for facts that may or may not be material in their ultimate nature but which are accompanied by consciousness and are not limited in either space or time. For instance, my body is material; I cannot move it except in space and subject to gravitation, and it has all the other characteristics of matter. But I can "think" of skipping from here to Shanghai and back without budging from my chair and I can think of doing it now, yesterday, or in the dim and nonexistent tomorrow. These are by no means adequate definitions of the difference between material and mental facts. They are, however, sufficient to show that the words can be used to distinguish these phases of what we call reality. Mental facts, thus understood, are somewhat limited by the physical structure, which is unequivocally material, and by the past experience which is both material and mental.

The second difficulty lurking in this dark place is the nature-nurture monster. We can refuse to challenge that difficulty by saying what all will admit, namely, that any human being, at any time, is the product of his structural capacity, that is, his inherited biological nature, modified by the experiences, the learning, through which he has gone. Nature and nurture are mixed up in him. For praise or blame, we cannot extricate the factors. In the nature, in the inherited capacities of any person, learning can make no change. But there are changes in the organism, taken as a whole, and learning makes them. In deciding what any person does with his possibilities, the organized learning we call education is one of the powerful factors. Hence, most teachers, out of understandable pride if for no other reason, are environmentalists, and all societies set up schools to see that environmental factors have the greatest possible effect.

One thing seems quite clear. In these uncertainties as to the differences in animal endowments, man against the whole menagerie, and in these uncertainties as to what is matter and what is mind, it is possible to see and to prove one decisive fact. Only man uses symbols to convey meanings, to communicate with others. This must be developed from an inherent, physically inherited, material-mental aptitude. Cats can grow in the same cradle with a baby but they will never prattle. Man can talk. He uses language and builds mental systems by which culture can be learned and in that way maintained. We can go from the physical to the mental by this one bridge, man's body can produce sounds and signs that other men's minds can understand. The beginning of humanity is the word.

IV

The "person" then, is our first entity. Scientifically defined persons are not real persons, unique and individual beings, because science can never comment on the unique. We are constructing for ourselves a working concept. *Let* the word "person" stand for the single human being considered only as a pattern of habits with a marginal capacity for change, held together by memory.

For sociology and for social engineering this definition will do; obviously we should have to describe a different concept if we were talking about persons or personalities from the standpoint of psychology, biology, poetry, religion, or love. Each mode of expression must abstract its own concepts from its own materials and set them up by definition, explicit or implied, before it can say anything important. Our sociology undertakes to make verifiable statements about the relations among persons in organized groups. The relations cannot be described until we have fixed the unit. A pattern of habits. Some power to change. Memory, conscious and unconscious, the binding of the pattern. This is the entity that acts.

We can try to avoid misunderstandings by saying that "habit," as here used, means "predictable behavior." Some disapproval clings to the word, however, and no matter how much we disavow any disapproving intention, some will dislike being called "just a bundle of habits." If they can be content, awaiting further explanation for a moment, to be known for a special purpose as a "bundle of predictable behavior," that will do as well.

Profound problems of being are skirted here. They cannot be solved, of course, nor even fully stated. We must leave them to the leisurely philosophers. But there is a moral and perhaps a scientific virtue in remembering, while making basic assertions about the person, that anything existing at all must exist by what it is not, as well as by what it is. Another way of saying it is this: Any predication is a denial as well as an affirmation. Plato,[4] struggling with the first emerging concepts of logic, had to show that predication is possible, in other words, that there is really such a thing as difference in the world. Two thousand years afterward, the Jew of Amsterdam [5] wrote to a friend, "It is plain that the whole of matter considered indefinitely can have no figure, and that figure can only exist in finite and determinate bodies . . . figure is nothing else than determination, and determination is negation."

Perhaps a biological parable will make the meaning, if not the importance of this statement clear. Life is a continuous chain of being, the biologist says, each living thing coming down from some living thing before it. The actual beginning is not a scientific question, for the present at least, since we have no

[4] Cf. especially in *The Sophist.*

[5] Spinoza, Letter to Jarig Jellis No. 50, in *The Philosophy of Benedict de Spinoza,* tr. H. M. Elwes (New York, Tudor, 1936), pp. 374-375.

methods of investigation that apply. We can guess; and men have made fascinating stories of the growth from the first green blob of chlorophyll, sucking up the energy of the sun, down to their lordly human selves. Whatever that story really may have been, there is enough evidence to show that simple animals have turned into complex ones. Creatures of one cell were followed, one new species after another, by more and more elaborate structures of many, many cells. Complexity makes differences between individuals more likely and more extensive.

And complexity makes for death. The single-celled paramecium, for instance, is only a slightly opaque spot in dirty water. Under the microscope, it turns and feeds. If you watch it long enough you will see it divide, and now there are two complete and vigorous organisms in place of one. Such a one-celled being is potentially immortal; it need not die. But neither can it be born. Has it a self? In some dim way, perhaps, for nature has less liking for absolutes than for leaps and vacant spaces. But its self is merged into the life of its kind; the parent of which it is a half; the progeny which it can bring to being only by its own disappearance. No death, no unique reality, only partial configurations on the stream. When the succeeding complicated structures of many cells are produced, then each such complicated creature does have unique and personal life. And each such creature dies. Death is a price we pay for being selves.

It would be logical to follow this brief aside with a next step into the religious concept of the person. Theology itself forbids the scientist to enter here and scientists willingly give over any wish to speak of souls. The concept of the soul is very old in human thinking; as far as we know, it is as old as humanity itself and in this sense, if in no other, every man who has been a man has had a soul. And the consciousness of some immortal aspect of our lives is deep in all our natural thinking; we cannot be so "scientific" as to leave it out. We can, for the moment, however, say that the soul is by its essence not subject to enumeration or measurement. If it is material and not divisible from the body, then we can call it the conative aspect of the mind, the same thing as the will. If it is outside the tangible world of force and matter, as all theologists have put it, then it is also outside any intellectual power the merely scientific thinker has with which to grasp its nature or describe it.

V

My own impression of my own personality or, in our terms, of the "person" that corresponds to me, is what I call my "self." We can get somewhat closer to the general concept of the person from the outside, however, before we look at ourselves. Predictable behavior is the action of the personality. Anything so strange and bizarre as to be quite unpredictable is either accidental or the result of a "personal" change. Predictable behavior is made up of habits, and habits are learned. They are acquired in experience by intention or by the purposeless but compelling tendency of living matter to repeat responses when stimuli repeat. Does that mean that the personality is learned?

It would be well to say directly that I have no faith in those theories of education that are based on an apparent belief that a child, at any age, is a sort of folded bud of all that he can ever be. Basic traits of the personality are acquired very early, no doubt; and they are not acquired by an undifferentiated abstract animal. A single unique structure, an embodied capacity, begins even before birth to organize the impacts of the outer world into reasons for behavior; and from that mere beginning the behavior has a pattern. We have learned something when we have taken on a likelihood of responding to certain stimuli or those sufficiently similar to them, in repetitious ways. If we do not repeat, our behavior is unpredictable and meaningless. A man does not then, strictly speaking, learn the traits of his personality. What he does is to become the person that comprises all the habits that are the repetitions which his experience has set up in his unique system of nerves.

Such a way of looking at the elusive facts of our individuality rubs out some of the crude distinctions that we make between human and brute self-consciousness. Some biologists, Hogben,[6] for example, seem to think that there are no distinctions except in degree along the continuum of reflex–instinct–habit–rational choice. The reflex is the mere unthinking jerk of a muscle when a nerve is touched, or the blink of an eyelid, or the workings of our busy viscera in the arcanum of the belly. The merest protoplasmic spots of animals have these.

Out at the end of the continuum that points toward freedom are choices not quite automatic, merely probable. Here is the margin of creativeness that is allowed by the not altogether rigid structure of the person. The predictability, however, not inventiveness, is the essence of the person[7] and the matrix of social relations.

What then becomes of our idea of "self." The term is another illustration of the general logical rule that definitions of the most familiar concepts are the most difficult. Persons who are the products and the carriers of Western culture, especially in modern times, find it very hard to imagine that the idea of self could ever be dim or multiple in a human mind. The anthropologists can prove that there are persons who actually feel, at least in childhood,[8] more group "we-ness" than personal "I-ness" and others in whom the whole idea is a shadow. In our highly individualistic culture the forces of language, discipline, affection, and family custom work in the child's first constructions of experience to make him orient his ideas around his "unique self." Being is not, for him, dispersed vaguely in the tribal group as it was for Todd's small Zulu.

[6] Lancelot Hogben, *Nature of Living Matter* (New York, Knopf, 1931).

[7] If I may refer again to Spinoza, I can point out that philosophically, also, habit, as I am using the term, is the essence of self-hood. Cf. *The Ethics*, Part III, Proposition VII: "The endeavor, wherewith everything endeavors to persist in its own being, is nothing else but the actual essence of the thing in question."

[8] A. J. Todd, "Primitive Notions of the Self," *Am. J. Psychol.*, XXVII, No. 2 (April, 1916), 171-202.

There are some confusions in educational practice. Enthusiasts of progressive methods seem often unable to decide which they want most, to develop a child's "individuality" or to "socialize" him into the group. Some of them are aware that these two purposes are in some measure contra-active and they have conjured a good deal of verbal magic to bridge the gap. A modern teacher may have an ambivalence towards the child's right to become a "self," but there is no ambiguity in the difference between social self and personal self as we are discussing it. All selves are both social and personal; we shall come to that later. The important fact here is that the very consciousness of self, in the sharply defined thinking of a modern, Western person is a construct. Social experience made me highly conscious of my me-ness. If I had been born in a culture where selves did not count for so much, similar but contra-acting experiences would have made me less self-conscious.

There is even a logical basis for this. Part of our self-consciousness is the result of a complex system of relationships between me and you, mine and yours, and so on. Some dull minds never attain full selfhood in this environment because the contacts are so impressive to them that their imaginations cannot make the abstraction we call a relation.[9]

VI

To say that the term "person" will be used here to designate a pattern of predictable behavior with a margin of creative change is intended to be an operational definition. It cannot perhaps suggest either the mechanical operations or the mathematical operations that usually accompany a definition in physics. It should be noted, however, that exactitude, even for the physicist, is an ideal toward which to strive and not a necessary condi-

[9] Cf. Elizabeth B. Lane, "Relativism in the Thinking of Subnormal Subjects as Measured by Certain of Piaget's Tests," *J. Genetic Psychol.*, LIV (1939), 107-118.

tion of successful endeavor.[10] Properly speaking, such a definition should be a list of things done in the mind, or in concrete manipulation, or in both, by which the entity defined is described, discovered, or measured. Bridgman uses the concept of length as an example of the fact that the physicist uses (in his strictly experimental work) only such concepts as are "synonymous with the corresponding set of operations." [11] The operations that are "synonymous" with the concept of the person are social in character. They are the operations of recognizing or identifying the human beings with whom we have some degree of close experience. The closer the intimacy of contact the greater the content of the concept of the person.

For example, we should expect anyone who looked like a human being of the Manhattan variety to get in and out of a subway car with us without too much trouble. These creatures against whose bodies we are pressed in an intimacy which is saved from being either immoral or disgusting only by the fact that it is colored by subdued hostility, these fellow travelers are persons in a very faint degree. One of them becomes a person in a richer sense if we see him often and note that he is always reading the New York *Times*, or always carrying a paper package, or always mumbling to himself. These become bits of predictable behavior more idiomatic, more "personal," than the crude amenities of subway society. An individual who turns out to be bound for the same office building as ourselves is still more of a person. And so on. What we are *doing* to create the concept in our own mind, by the organization of our own experience, operationally, is to tie together a more and more complex and subtly related set of repetitious traits of behavior until, as we say, we know and recognize the person. This means that we act in the confident expectation that he will act in certain ways.

[10] P. W. Bridgman, *The Logic of Modern Physics* (New York, Macmilland, 1932).
[11] *Ibid.*, p. 5.

The physical appearance, as we have said, is only a symbol of the self and we are quick to reject it if the behavior disappoints.

There is, moreover, the margin of expected change in the pattern. It is not a flaw in the predictability of repetition but itself a form of predictability. It is not merely the "penumbra of uncertainty" that hangs over all operational definitions. Bridgman says "our concepts are not well defined things, but they are hazy and do not fit nature exactly." [12] The physicist is always, in his philosophical moments, insistent on this tentativeness and uncertainty which makes unnecessary if not downright unjust the philosopher's constant reminder that the scientist does not depict "reality," or Whitehead's "Exactness is a fake." [13]

We are putting into our definition of the person a factor of change, along with the factors of uniformity, because it is unimportant to say whether this part of the person is determined by past events or is "free." Since these actions are free in our operational situation, that is, since they are a part of the process of recognition and social reaction, they are free as part of the concept. The philosophic and religious problem of the free will does not now arise.

There is, however, what happens to be an almost inescapable philosophic question here that does affect our thinking. I believe that it is also a scientific question. The question is: Are the unpredicted actions of persons real changes, not in the sense that they have no causes in antecedent events, but in the sense that they are the first observable phenomena of a series that will develop under observation into modifications of the predictable pattern and hence eventually of the social relations of the person? The answer is that they are changes in this sense because this is the same as saying that the locus of social change is in the person, a point that seems obvious but is actually of great theoretical importance.

[12] *Ibid.*, p. 201.

[13] "Ingersoll Lecture on Immortality," *The Philosophy of Alfred North Whitehead*, ed. Paul Arthur Schilpp (Evanston, Ill., Northwestern Univ., 1941), pp. 682-700.

The whole concept fades out in some kinds of thinking; it ceases to be useful or meaningful just as the concept of temperature fades out when the thinker gets to the atomic scale of magnitude.[14] In operational thinking all concepts are restricted to the field and the scale for which they were devised. This is important because it raises no questions as to their validity out of their field, as the concept of temperature is not made less useful or valid in vast ranges of thought by the fact that atoms are neither hot nor cold.

If we still balk at accepting, even for working purposes, a definition of the person as essentially a pattern of habits, it is easier to see its truth in applying it to our friends to whose feeling of self we have no access. What after all is the "personality" of a friend? How do we know the person with whom we have pleasant enduring relationships? It is by the pattern of habits we have learned to identify with a feeling of "friendliness." It is by a pattern of confidently expected, that is, of "predictable" behavior. If a friend behaves in unexpected ways we are disturbed, more or less, depending on the degree of surprise. If a person looks and speaks like someone we know but in no other respect fulfills our expectations we are apt to conclude that it is "really somebody else," and indeed it is somebody else, because the mere physical body's apearance is not our friend in any sense real or emotional. If the divergence is slight but still outside the range of variation that we tolerate in our companions we say, "You are not yourself today." This means, of course, "You are not acting entirely within the pattern of habits that is essentially you."

Split personalities are literally multiple persons in a single body. Each of these persons has a pattern of habits but you cannot be friends or have psychological relations with a group. You can only take them one at a time, person by person.

The literary uses of the word "self" have given us a logical structure, a habit of speech, which makes it somewhat difficult

[14] Bridgman, *op. cit.*, p. 224.

to use the word as we are using it here. We say a man "has," that is, "possesses" a fine personality, a pleasant nature, a strong body, or a keen mind. What is the real subject of that verb, "has?" The man, himself, is the common-sense answer. But the man himself, by any strict thinking is nothing more (or less) than the sum of all the parts that we are talking about. The man is the sum of his physical inheritance and all the things that have happened to him since he was born, his accidents and his learnings.

Is this mysterious possessor of the man's characteristics his immortal soul, then, and is the soul the owner to whose benefit the debits and the credits must accrue? This is no doubt true in what we call a spiritual sense. That appears to be what the theologians mean by the term "immortal soul," and we have no reason for using it except as they do. But these last two sentences of mine are meaningliness in the world of science, the world of abstract uniformities which can be logically constructed out of the evidence of the senses. In that world the possessory, active way of speaking of the self is meaningless except as an illogical idiom. The self does not *possess* all these things, its attributes; it *is* its attributes. Or, to put it in other words, a man does not have a good body, a mind and a disposition; he *is* those things. In a world of values, to be sure, the whole is often greater than the sum of its parts. You cannot create a man for philosophical purposes by merely adding qualities, nor find his essence by tearing him apart. But Spinoza said the essence of self is the persistence of the pattern. The habitualness or predictability of behavior, the action aspect of the attributes we have been enumerating, is for every man the tangible entity, the scientific "thing." This thing we call the person.

So, we do not here say that a man is "conditioned" by the culture in which he lives. His pattern of habits, his personality, is a set of choices from among the possibilities which the culture offers him. His notions of self are some that his experience has taught him to prefer from among those current in his time and place. There may be inescapable implications of self in mere consciousness but they are inaccessible apart from cultural expression. In this they are like all other human qualities; we can never find nor observe any human behavior outside a cultural frame.

What then of personal differences that seem to be impervious to nurture? What of invention? Of freedom? Above all, what of freedom? It is not really a digression to point out that there is no logical inconsistency between the scientist's ideas of cause and effect and the idea of freedom. It is possible to believe in cause and effect as a logical truth if not a scientific one. Consider the possible logical choices, imagining for the moment that your illusion of freedom of the mind is not mere self-deceiving. If the world has no uniformities, if all systematic description is impossible because the universe has no order whatsoever, then there can be, of course, no freedom, no choice, because no event has any relation to any other.

On the other hand, if the universe is a seamless web of necessity, then cause and effect are meaningless terms because every event, every thing that happens in time and space, is both cause and effect of everything else. No one, I think, was ever such a determined determinist that he thus reasoned cause and effect right out of the world. The hardiest deniers of freedom are usually those who see small tangible patterns of necessity afloat in vague contingency. They ascribe a kind of secondary reality to the matrix of chance and say that it too can be brought into the rigid structure when it is better known.

For our purposes, we repeat, freedom is real in the sense that in the person there are experiences that involve uncertainties, preferences, and acts of choice. We know this from observation of others and of ourselves. They have experiences in which they see consequences in events which they connect backward with actions of their own. They feel pleasure if these events are good and chagrin if otherwise. This provides us with the reality of freedom in our own experience and that is what we are talking about.

There is more, however. Reasoning can show that the kind of self we are talking about is free. When a choice is made, the dominant elements in the personality have made it. The uncertainties have been possible because some elements of the person have found good in one possible course of action, other elements in other possibilities. Uncertainty is a conflict. I am coerced! It is not my will but my need! I did what *I* did not want to do! These cries, emotionally sincere as they may be, are likely to be confusions of words. Choice is possible in only one way and all action follows choice, conscious or inherent. When it is clear that some factors outweigh others, then the self chooses, and if conflict remains that is because the self is divided and has more than one desire. Nothing external has imposed the conflict and only internal adjustment can bring harmony to the warring parts.

INSTITUTIONS

I

IF WE WERE DETERMINED TO THINK SCIENTIFICALLY ABOUT HUMAN behavior we would thus make as much progress as we could toward reducing the primary units, such as the person, to univocal precision. We would try to define for ourselves the entities about which quantitative generalizations might be made. And we would here use the pattern of predictable behavior as a first unit of thought because it is operationally expressive of the relation between the person and the group in which the person carries on his social life.

Each person comprises a number of habit patterns that are shared with the habit patterns of other persons in a complementary fashion. This is what makes group life possible. And we can define another primary entity in these terms. A set of habits, shared by two or more persons in complementary relations, a predictable relation among the predictable behaviors of two or more persons, will here be called an institution. Culture is a pattern of institutions. They are the kind of knowledge or mental content that enables you to know what the other fellow is going to do next. They are the substance of that complex of expectations that we need to support any kind of social action. As far as we know, man never did live in desolate aloneness. In that loneliness he would not be man. Something intermediate, a gregarious half-man, between the solitary animal and man, is necessary at least as hypothesis.

Thomas Hobbes was probably all wrong about the "covenant" that was "the Generation of that great LEVIATHAN, or rather (to speake more reverently) of that *Mortall God*, to which wee owe under the *Immortall God*, our peace and defence." The state came, it seems, out of the adjustments of repeated action, not out of compact. But Hobbes understood well enough that man is man only in society and he put it in dour eloquence,

> Whatsoever therefore is consequent to a time of Warre, where every man is Enemy to every man; the same is consequent to the time, wherein men live without other security, than what their own strength, and their own invention shall furnish them withall. In such condition, there is no place for Industry; because the fruit thereof is uncertain: and consequently no Culture of the Earth, no Navigation, nor use of the commodities that may be imported by Sea; no commodious Building; no Instruments of moving, and removing such things as require much force; no Knowledge of the face of the Earth; no account of Time, no Arts; no Letters; no Society; and which is worst of all, continuall feare, and danger of violent death; And the life of man, solitary, poore, nasty, brutish, and short.[1]

We are not trying now to discover the origin of the social patterns. We are experimenting with the task of describing them in terms that will make operational thinking possible. The human being, organizing his thinking, his naïve aggregates of remembered experiences by means of a set of symbols, is a specific pattern of inherited animal characters which, as it grows, becomes a specific pattern of choices that harden into habits made out of the alternatives offered him by his culture, in his time and place. Institutions then are embedded in the habit patterns of persons; they are mental. To go with them, there is another group accumulation that grows from generation to generation, the heap of tangible things which some have called "material culture" but which I call "tools."

To put it in another way, the habits of persons are shared with other persons in reactive fashion, and these patterns of predictability, existing in the relations among persons, are institutions. They have material concomitants and these, again defining scientifically in the imperative mode, are tools. And culture is a com-

[1] Thomas Hobbes, *Leviathan* (Everyman, 1913), Part I, Chap. 13, pp. 64-65.

plex of institutions, or mental patterns, which indicate to men the uses of their tools.

The nature of an institution may be illustrated in a fable about the food hunt. My own behavior in search of food is, as we say, much "sophisticated" when compared with the pig-stalking of my Cro-Magnon ancestors. It lacks direct intention. I do not sidle down the walls of a New York street toward a grocer's unguarded bank of fruits and vegetables. Nor do I, in spite of opportunity, snatch a handful of victuals and scamper to a dark corner where I can chew and growl.

I behave in a seemingly disinterested fashion, expressing my "self" in a set of habits, whose relevance a primitive could never guess at and whose successful outcome depends wholly on cooperative habits of other persons. I work mostly with words, spoken or written, pronounced in a classroom or toward a microphone or taken down by stenotype. Those words have power; they determine the behavior of others in more or less cooperative ways, but since I am a teacher they also eventuate in pay. Slips of paper are sent to a bank and I send slips of paper to grocers and butchers and at the end of the cycle that began with a sign or a sound I am provided with a dinner.

The only reason for going over this very well-known cycle thus in detail is to enforce the fact that it is a food hunt as well as, I hope, some other things also, and to call attention to the fact that in its working my habits are geared into the habits of students and clerks and engineers and broadcasters and bankers and grocers, and so on and so forth, far beyond any knowledge I need to have of processes or persons. And these geared-in shared habits are sets of institutions. The mere thinking of all this, or dreaming it, would not keep me alive. The institutions, the relations of adjustment and cooperation, the mental patterns, must have controlled materials to work with. These materials are tools.

A church, for example, is an "institution" in the looser sense. In our more exact sense a church is a complex of both institutions and tools. A church has ritual and worship, art and tithing and an overt creed, and these are all institutions. They are co-operative relations or sharing among the habits of persons, and are organized in their minds by means of symbols. So also the church has tools. It has temples and altars and vestments, strong boxes, musical instruments and books. And in the mind of every person there is a feeling of permanent relationship to his "church" or his "religion" which is vaguely but sufficiently attached to images, and there are emotional tones that float about the tools and the institutions both, and require no definition.

II

We are now thinking of persons as patterns of habits and of social groups as patterns of interaction by means of habits that are shared. The person has, of course, many habit patterns that are more or less peculiar to himself; they make his individuality or uniqueness. His own version or reciprocal part of the habits he shares with others is also peculiar. It has often been said that he has in fact a set of different selves, each of which is used when he is acting in the context of some one of the institutions in which he takes part. The idea was not unknown to Adam Smith and has fascinated modern analysts like Charles Horton Cooley. I have preferred to label the combination of all these phases the self, more technically the person, a complicated entity.

For our own manipulative purpose here, the important fact is that the various social relations of the person involve a large part of his emotional life. With the pattern of overt action there is a concurrent pattern of loyalties. A loyalty is an habitual or predictable and relatively enduring emotional tone of approval which attaches the elements of our self-respect to our opportunities for self-expression.

If we are going to influence people toward social change, we must be able to describe with effective accuracy the emotional elements in their social behavior. Our loyalties, our general enduring approvals, begin with the attitude toward the self. Loyalty to the self is an over-all pervading emotional factor that resists all change.

At this stage of our argument, our distributed loyalties are more important because loyalties to institutions are the emotional cement of social structure. They may go deeper than emotional approval of friends because we sometimes cling to institutional relations against the wishes of our friends, as when we give up a friend rather than a political or a religious allegiance, and we often feel a strong love of an institution without having any person in whom, for our imaginations, the institution is embodied. In fact, a democratic, or pluralistic society is one in which institutional loyalties and personal loyalties can be separate.

Every person is a unique pattern of habits and these habits mesh into the habits of other persons, making institutions; as a result in every person there is a unique pattern of emotional attitudes that are the attachments of loyalty felt toward the institutions. The more complex the society, the more varied the loyalty patterns in different persons. The same thing can be said of the size of the social unit. But in a modern society, even in a small village, each person will have a pattern that differs materially from the loyalties of his neighbors. The carpenters of the town will probably not all go to the same church; that is, their occupational and religious loyalties will differ. Nor will all the lawyers belong to the same lodge nor all the members of the golf club be Republicans. It is easy to see some standard combinations, of course, but it may be said that democracy is best developed in those places where the pluralism of patterns is taken for granted. Men differ. In a democratic society they enjoy each other's differences. And a cultural group that can allow comfortably for difference can change without great cost.

EDUCATION

I

SUPPOSING THAT WE HAD ADEQUATE ENTITIES FOR SCIENTIFIC PURposes, the person, the institution, the loyalty, what would the social scientist do with them? What would the theorist do? The engineer? Sufficient answers to such entirely proper questions cannot be given here, even if most of what has already been said is accepted. But we can exemplify the logic.

The use of such entities in theoretical description, in the first phase of science, is not so unfamiliar nor so difficult. In fact, almost any statistical statement about the behavior of human beings in society is a statement about units that are patterns of habits. Whatever else there "really is" in the concrete existence of the individuals whose behavior is summarized is left out of the proposition. The statements may be positive. We might, for example, say that 5,310,608 inhabitants of New York City ride on the subways, on the average, every day. This says something quantitatively about a class of habit patterns. It is positive because it describes predictable continuing behavior. If we should say that such and such a percentage of the inhabitants of New York would commit suicide in May of each year, we would be making equally a statement about predictable behaviors, but we can call it negative because it is a prediction of reversals in the habit patterns of single persons that will maintain a continuing pattern only in the statistically described mass.

Further, we might say that a certain proportion of the inhabitants of New York belong to and go to the Presbyterian churches. This is a more important kind of a generalization for the engineer who intends to use the theoretical descriptions, since it implies statements about loyalties. To know that 5,310,608 New Yorkers take the subway does not help much in knowing how to change the behavior of New Yorkers in any direction. It would not, for example, help us much in our quest for the ways in which culture can be changed to enlarge freedom. But the statement as to churchgoing would be of use. The engineer of culture, the one who sets out deliberately to change human habits, and institutions as social phases of habits, would have to seek always the relevant facts about the loyalties, the emotional trends of his material.

The builder of a dam might say, "I can count on this measured quantity of stability in this south anchor of my dam. I know that I can rest a weight of a measured number of tons on that anchor and it will serve my purpose. I can build a trustworthy dam."

The breeder, using the theoretical findings of biology might say, "If I cross this breed of cattle with that one, I can accomplish my purpose. I can get a strain of cattle that will produce more beef."

If the social engineer could work on a basis of equal certainty he would want to make statements like these about human loyalties, that is, motivations, among which he could intervene in order to accomplish his purpose, which is the changing of human behavior. It may be necessary here to point out that I have been, all along, using the term engineer to designate only the one who would advance social change, not the one who would instigate or direct action within the existing patterns.

Let me illustrate. A leader like Hitler, for example, may bring on a catastrophe by the manipulation of human behavior. He may even lead men to believe that he is making deep and far-reaching reforms in the cultural patterns while he is actually only carrying out implicit phases of the old pattern and making no change whatever. He may even push the old pattern, as Hitler pushed the German domination ideal, to the edge of destruction or even wholly destroy it. He is still not, in our meaning of the term, a social engineer. A social engineer working in Germany would be one who would take scientifically precise descriptions of the present patterns of institutions and loyalties and would intervene to change them, as one might, for example, bring a new pattern of democratic give-and-take into the German complex instead of the present status anxiety.[1]

But the social engineer and the charismatic demagogue would work with the same instruments. They would work with symbols, because culture is a dynamic organization of symbols in the minds of living men. To bring men into action in any implied or customary phase of their present patterns or to lead them into change, you would work in the same way, by intervening in their lives in such a way as to change the flow of symbols through their minds. This is only a more elaborate way of saying that you would persuade them and you would educate them.

The person is the pattern of habits—but with a marginal capacity for change without losing identity. The social engineer would consider that the building up of the original habit pattern—education—was part of his business as well as the influencing of the older patterns by persuasion.

In education, the social engineer would seek to build in as many persons as possible, in the "normal" person, the habits of free action. Still more, he would try to build loyalties to the institutions of freedom. In moral terms we would say that he would try to build in the normal person a love of the freedom of others. It is probably a verifiable proposition that freedom exists in any social group in the degree that the members of the group are concerned about the freedom of others while they exercise their own without hesitation—but especially that they are anxious to maintain the freedom of others. But first we need to place the set of institutions we call education in its place in our concept of culture.

[1] "Germany after the War: Round Table, 1945," *Am. J. Orthopsych.*, XV, No. 3 (July, 1945).

II

Culture is the organization of the energy that human beings exert in social living in comparatively stable patterns of habitual behavior. In so far as it is a thing and has a locus, it exists in the minds of the persons who inherited it, are busy living by it and changing it. The persons who are living by their own cultural patterns, or more accurately, in whose habitual behavior the patterns of their culture can be observed, are not aware of all their motives and may be quite often mistaken about them. But their actions are not on that account merely reflective or instinctive; they are still purposeful. Speculation about the sophistications of those purposes and the gaps between intention and self-understanding make up a good deal of the sociology of some iconoclasts, like Pareto, but the fact that men have always fooled themselves was evident to Eve. The important point for us is that the continuing basis of culture exists only in the minds of persons, and if we want to make deliberate changes we get at them by changing men's minds.

The most elaborate institutions that any culture has for changing and molding men's minds, giving them initial shape and then reforming them in time, make up what we call education. Culture is not a physical organism in any real sense, but it has nevertheless this characteristic of living beings, a power to continue itself, to reproduce. The inheritances are direct, person to person, but there is a division of labor and some persons are dedicated to the task of establishing the inheritance. Education is our organized ways of learning from the experience of others.

A social engineer, then, would always be in some degree an educator, although he might not be a teacher in a classroom. The institutions of education would be one of his primary concerns. Here again, he is not undertaking to set the purposes of education. For our discussion, we postulate the value, freedom, and go on to indicate, if we can, how education can help to form persons whose loyalties would be attached to the institutions that make freedom possible. We want to know how to educate men who will be happy only when they are free and who know how to live in liberty and how to keep it. This is a standard Aristotelian idea; we want to know how to give them the habits of free men.

Education itself is not a science. It is a set of institutions, of patterns of habits, shared by groups of persons. It is an "idea," if you like, and it comprises certain arts, such as teaching by spoken or written words or by other symbols, as well as complementary learning, and research after new knowledge. The institutions make use of tools, material possessions that are found in nearly all modern societies in greater or less elaboration, buildings, books, chalk and ink and accumulated dust. The idea embraces also certain aspects of some of the sciences, mainly psychology, to tell us how teachers and learners behave, and sociology to tell us the meaning of this set of human relationships. It is also, of course, an ideal and an idolatry, a fiercely self-protective professional group and a universal experience. "Education," the word, can mean any or all of these things and many more. And any one of its meanings can be used with positive or negative passionate reactions. It is a miracle that two people can ever talk about education and even suspect that they are talking about the same thing.

We can begin with a provisional definition of education as institutional behavior that provides opportunities for purposeful learning. Several of the words in this sentence must have special and arbitrary meanings and the definition cannot mean much to begin with. But it seems best to start with something that will have meaning at the latter end of our discussion rather than to offer an obvious tautology and then have to give it meaning by qualifying, bit by bit.

I exclude from our meaning of the term education, for our purposes here, all those definitions that make education merely another word for experience or life. If we were beginning biologically, we could take off with the statement that everything that lives, learns. Life and power to learn are almost but not quite the same thing. Hence, life and education—! Specious statements of that kind are all about us. Naturally, all experience teaches us in the degree that we are capable of learning. But life and experience are not subjects, in this vast syncretic form, that anyone can usefully talk or think about. They are infinite aspects of existence and we are finite minds. For good methodological reasons, we can leave poetic definitions of this kind to those who are amused by them and talk about something more precisely limited. We want to use the word to mean something in terms of possible action. We deliberately narrow it to designate some phases of experience or of life that we can grasp intellectually and change practically, if we see reasons for change.

By calling it "institutional" we mean, in our own terms, behavior that is part of the habitual behavior patterns of a number of persons. And by limiting "learning" to what is "purposeful" we are cutting our concept down to manageable size. As a matter of fact, what is so laboriously outlined here is the common-sense meaning of our term. The ordinary citizen when he says, "education," does not mean "life"; he means learning something on purpose.

This purposeful activity takes place in all social groups, cultures, or nations. Without it there could be no social behavior; man would be the solitary ape. It is not always elaborate, of course, but it is always purposeful. Its purpose is to maintain for this generation a part at least of what other generations have learned. Secondarily, it has the purpose of criticizing culture and changing it. In its high ranges, it has also the purpose of freeing men from slavery to the cultural necessities by which they live. It is in this area that we are likely again to have trouble with the word that we have been using from the beginning, the word "culture." As Hogben says:

> In theory, the word cultural commonly covers two entirely different functions of an educational system. One is the private

> problem of helping the individual to discover for himself or herself congenial sources of enjoyment to occupy leisure in later life with the fullest allowance for variety of temperament. The other is the public business of equipping individuals with the knowledge necessary for the discharge of their mutual responsibilities as co-citizens of a democratic society without regard to the personal inclination of the child.[2]

In this book we are speaking of culture in what might be distinguished as still a third way, as the word is used by sociologists rather than by essayists on matters of education or on general ideas. The quotation from Hogben is given so that we can clear off those two possibilities. But what we have been using the word to cover is all that is essentially human, all that is accumulated by man from generation to generation, all the ideas, languages, skills, sciences, lies and mistakes, faiths and virtues, all the equipment for living that can be in a man's mental possession. We can also use the word distributively and speak of "cultures" because each group has its own inherited possessions in mental patterns, different more or less from all the others.

The social purposes of education are to keep these "cultures" in existence. The purpose of one educational system is to keep one pattern of these things in continuous life. Then to criticize and change it without breaking continuity. Then to free as many men as possible from its negative compulsions.

Persons, individuals, single human beings are the only vehicles of "social" purposes, as they are the only possible vehicles of any social fact. They may have private purposes, also, however. They may wish to move in the social structure and may seek a superior knowledge of the culture for that gainful activity. Or they may have other private reasons. We shall confine our attention to the social purposes.

Education is the means by which social groups carry on their culture and revise it as they go. How the process of school-

[2] Lancelot Hogben, *Dangerous Thoughts* (New York, Norton, 1940), p. 108. Reprinted by permission of the publisher.

ing the young is related to the carrying on of institutional patterns has already been discussed. It would be impossible to have a culture of any kind, even the meagerest sort, without some kind of education, since a culture is a continuing pattern of human behavior which gets its continuity from the fact that children can learn from their elders, and so, in an unbroken chain of minds, the pattern is learned and lived by. As we have tried to say, man's advantage over the rest of the animal world, internally provided for by his greater learning power and his lesser instinctive fixity as well as by his long dependent-infant period, is outwardly implemented in the systems of symbols by which he maintains continuous, mutually understood patterns of behavior. A shorter way to say this is that man learns from the experience of other men.

We need not spend much time in discussing this "transfer of culture" which is the first and the indispensable job assigned by a social group to its educational agents and institutions. If culture is not transferred, culture ceases. In this sense education must be derived from *educare*, to rear or feed, rather than from *educere*, to lead out, not only because the processes feed the growing appetities of the developing person; they are the alimentary system of culture itself, as well as the procreative.

A few modern educators, strangely afflicted with misunderstandings of Dewey, have seemed to think of trying to educate a child entirely out of his own experience instead of by giving him the benefit of the experience of others. But even they have betrayed their own imagined principles, because in directing that the child should learn chiefly by his own experience they have drastically manipulated the circumstances of the child's experience in a way that made the teacher's previous experience a major element in what happened to the child. They chose his experiences for him, deciding in advance what he ought to go through. Even the judgment, man learns by free experience, is a judgment contributed by the past experience of other men. It seems unlikely, however, that there will ever be more than transitory vagaries of this kind. Transfer of culture remains the natural, necessary, and easiest phase of organized learning.

The second phase, critical revision, is more difficult and there will always be some who will, from piety or self-interest, stand in the way of the schools when they encourage change. A satirist of education may some day make a major comic character of the mature man who remembers a highly admirable self, which he thinks was his own self in his school days, and who wants the school to make over his children in his own imagined image. He never was what he thinks he was, of course, nor are they what he thinks they are. But he, nevertheless, is a bar to progress because he wants the old frames preserved, the subjects of study, the hours of work, the methods of instruction, those all must be as he remembers them because he remembers with such affection the small person he thinks existed once and which he thinks those methods helped to make into something wonderful.

I once took the trouble, in a fit of annoyance with the criticisms of the schools in a Middle West American town that were made by a leading merchant of the city, to look back to see what the businessmen of twenty-five years before had said of the schools when that same businessman was graduated from the local high school. When I quoted to him the phrases, almost his own, spoken by the businessmen of his father's day against himself and his companions as children, he was chagrined. Not silenced, of course, but puzzled. He could not believe that the slow degeneration of the school system was not a phenomenon of his own time. He was sure that he had been one of the first to notice it. And if the businessmen of his high school days thought boys like the boy he was were careless, underdone little rascals, what, in the name of heaven, he asked, would they think of the boys of today! That the boys of today are very much like the boys of yesterday and that the schools have changed mostly for the better is difficult to prove and, as far as citizens of this type are concerned, never to be believed. And this is part, of course, of an ironic habit common to all the generations in their mature years. They urge the young to make the world over and make it better, but they cry out in agony over every change.

The critical evaluation of the culture by the school is not quite the same thing as the problem of the responsibility of the schools for a "new social order." The question asked by George S. Counts, "Can the schools build a new social order?" may have had some meaning as Counts put it a dozen years ago. Most of the discussion since has been manifestly unreal. The schools, as an instrument of society can, of course, do only what society wishes or will allow. Teachers may have a greater chance than other citizens to mold the future; that is a debatable point. There is very little in the record to show that teachers, who are servants of the public, have ever been very successful in making any social changes the public did not want. This, however, has to do with what seem to me quite superficial aspects of the teacher's power. If the question means, Can teachers change the overt forms of social behavior, without having help from some other part of society? the answer is, No.

In a deeper sense, however, teachers may make more changes in the social order than are made by any other conscious agents. Certainly statesmen have no such power over the future. As long as ideas are to politicians the counters of power (*vide* Pareto) and are still, to teachers, real as values, then the teachers can inculcate values in the wishes of the people and the politicians will have to fulfill those wishes. They will be compelled to use those values and ideas as counters in their own game. This needs more discussion in another place. What we need to say here is that organized education exerts its influence on the future of the culture by the ideals and the loyalties that it builds into the persons of the oncoming members of the group.

In this way then, education as a set of institutions is a means of criticizing the culture itself. If we can succeed in understanding social change as the group aspect of individual change, we can better understand that the wide range of difference among members of the same group is a chance for rational creativeness. For many reasons, the teacher is likely to differ in his ideals from important contemporaries. He is ("she is," would be closer to the picture of the public schools in America) often an educated person. Sometimes he has himself been made a rational critic of what he teaches, knows the skeptics and creators of the past, and has the sense of professional responsibility which, even in a reactionary, insists on what is "best" in the cultural heritage. The teacher may be thoughtful. He is probably unlike the man of action who thinks in rough categories and is for that reason impatient of the qualifications out of which effective criticisms are made. Men of action are not likely to be teachers although, alas! they may be school administrators or members of the board.

Teachers will be more effective critics in the future, no doubt, when the social status of the teacher is enhanced. The overworked teacher of a rural center, paid less than the farm laborer, remote from professional guidance, and watched too often by a bigoted school board, may take heart. It is one of the most notable characteristics of advancing civilizations that the servants of the people gain in power. The barber surgeon becomes the physician and the public health worker, and saves our lives and is honored for it. The feudal chaplain that once ate with the servants in the kitchen is now the honored priest. Of course, there have always been bishops with power and professors with pretentiousness. The point here is that occupations as a whole change their status, and the general change appears to be for those that are most useful to men to gain more and more in prestige and in chances to be effective. It takes some courage to say that men who are destructive of cultural values, men of violence and chicanery, have less and less prestige as civilization goes on. There are some reasons for thinking that the change affects them too. In any case, rising status of the teacher is evident, and it is the ordinary trained professional, the ordinary person who is getting greater social power, not the prelate nor the occasional great one. Such a change is inevitable because the increase in all kinds of knowledge makes for constantly increasing division of labor, as we have already said. Such specialization makes each social role unique and indispensable. The possessor of special knowledge has special power.

In a summary it is difficult to deal with the third function, enlightenment. The word is arbitrarily chosen. The quality described might as well be called the lack of cultural provincialism, or we might say that the third general function of education was to give every member of the group a lengthy "ethical radius." Boas has most clearly pointed out what all anthropologists seem to believe, that time works for peace and stability, if men will meet one another on any peaceful ground.

The most primitive peoples are expected to be good to each other. They do not steal from each other, or commit murder within the tribe without being condemned and punished. And tribesmen hang together and defeat their enemies largely because the ethical codes work within the group. But on this primitive level, it is no sin and no crime to murder, cheat, or rape the enemy. Acts of violence against outsiders are acts of tribal power. It does not need saying that modern forms of nationalism are equally simple in their division between good and evil—good is doing good to the members of the tribe and evil to all others, and evil is the opposite.

Civilization goes forward as this kind of nationalism dies. Our discussion of the problem of war should have made it clear that wars will go on as long as the concept of "enemy" can be put in the place of the concept "human being." An education of complete effectiveness, of ideal power, would destroy the "enemy" concept in all members of all nations.

The process is subtle and difficult to manage. We inculcate loyalties to the culture by which a child must live if he is to live humanly at all. These are not merely "basic" to his personality; they are actual elements of the "person," his pattern of predictable behavior. And yet, there is danger to others if he is too blindly loyal to the elements of his own "self." Some persons may rise to a level of understanding that will make them know that although they must live inside some one of the cultures that exist, must probably live in the one to which they were born, they can still see values in others. They can see that the chief reasons for choosing the one they live by are that it was the one available to them, and that many parts of it are desirable in the sense that some kind of a solution is necessary for institutional problems and one is more or less as good as another.[3] If he is lucky enough to have been born in a pluralistic culture, he can still choose, on his own temperamental scale, many elements in his own pattern and keep both his loyalty and his sense of relativity. He can be more easily enlightened; and it is in such a culture, where patterns or normality are numerous and diverse, that enlightenment will be more common.

Enlightenment, as freedom from the cruelty and slavery of provincialism, is in one aspect a training of the emotions. Some of the deepest emotions are the dangerous ones. They are dangerous to mankind as a whole because they go along with an honest loyalty to what is believed good for the group. Notable in this is patriotism, loyalty to one's nation. It is very difficult to make clear to a young person, or to most older persons, that patriotism may be valuable to both the group and the world if it is mild and reasonable, but definitely dangerous to both if too intense. What kind of virtue is this, if one can have too much of it? The golden mean of Aristotle, the temperance of Plato, make a difficult rule to live by. To instill this kind of self-control, this denial of excess of virtue in the general run of humanity, is the most difficult task in education and is not generally successful. Wide knowledge brings moderation and healthy self-suspicion to a few fortunate beings, but most of us can learn it only as we learn any other social good, by being taught it in the experience that is specially devised for that purpose, that is in

[3] Pascal, *Pensées*, Part I, Art. IX. XLIII: "la coutume doit être suivie dès là qu'elle est coutume, et qu'on la trouve établie."

education. Our general failure to accomplish this result in education is shown in the good men who are always ready to be cruel in the best of causes.

Enlightenment is moderation in our loyalty to the culture by which we live, not because the culture is unworthy of loyalty but because there is positive value to humanity as a whole in a multiplicity of cultures in the world. An organized diversity inside any culture is beneficent; likewise in the mosaic of patterns that make the secular life of humanity there is a value that can be preserved only if we are willing to let alien cultures live.

III

The philosophers of value have never, it seems to me, come to grips with the problem of those virtues that are evils if seen from the other side. No ethics for humankind can ever be possible unless the chief virtues in any nation can be practiced without harm to good men in any other. And it may be found, if ever such a philosophic problem is solved, that we need not abandon any virtue but only to temper them all. Evil lies in excess, not in the choice of action. We are far from believing this. We still believe that good can be achieved by doing evil—if we do enough of it, and we are far from seeing that even too much good may be not right but wrong.

The choice of evil as the tool for great purposes is common enough in statesmen, even when they are saints as well as rulers. The counselor of Richelieu, *Eminence Grise* [4] is an instance. The Jesuit hero was not merely ambivalent because he had two ambitions, one of which was discipline for himself, and the other greatness for his country. The tragedy of such a life lies in the fact that the hero is led by the very strength of religious conviction, which like any other conviction can be evil if held excessively, to kill men and starve children and ruin empires to fulfill his own notion of God's will.

I have said, and have been scorned for saying it, that a perfect

[4] Now recalled to us in *Grey Eminence*, by Aldous Huxley.

society would be one in which no one would die for anything, not for any idea nor any ideal nor any good. It is an assertion that needs further development. It is not to say that men in this imperfect society we have now should never die for what they think more valuable than life. But the evidence on our wartime behavior shows how the impulse to die rather than give up a desirable good is separated by a mere shadow line from the impulse to kill. A saint may possibly be wholly passive in martyrdom, hoping to make conviction real in other men's minds. The story by Maupassant, *La Mère Sauvage*, is the more conclusive comment on mankind. You will remember the old woman who loved the Prussian boys quartered in her farmhouse because they were so much like her own boy who was fighting other Prussians off in the north. But when her own boy's death was reported to her, she killed the red-cheeked German boys with calm cruelty. One who feels the tragic blow of a friend's death in battle can feel soon after, without noting that he has crossed a line into another region of character, that he must kill some enemy to pay off the debt. These are tangled and little-understood skeins in our nature. As long as violence is the only method we can devise for stopping evil behavior, we shall continue to have violence and evil behavior.

What I mean to say here is not merely to repeat in these lifeless words what has been burned into the memories of men by all the heroes of religious faith. The point for today is that the failure is of intelligence, not of good will. We are constantly failing to break the long chain of wrongs, each one of which is revenge for wrong done before, because we are not intelligent enough to find other ways of solving the problems that lead to conflict. It should be one of the chief purposes of education, as it is of statesmanship, so to manage situations as to keep them from developing into conflicts.[5] But, as we have said, education must do its work in order that statesmanship will have an opportunity.

[5] H. D. Lasswell, *Psychopathology and Politics* (Univ. of Chicago Press, 1930), Chap. XI.

Statesmen come into the game after the counters have been distributed. They can hold power, in any kind of government, only by helping citizens to achieve in some measure the values of their culture, the values their people have learned to want. A people trained in moderation will not have the immoderate degree of patriotism that can be elevated into a doctrine of a dominant race or of military strength nor a doctrine of personal sacrifice for imperial glory.

An enlightened people, in which there was a true but moderate loyalty to the culture as a whole, would be a people whose attachments to other values would be also temperate and wise. This may seem, to those who have not pursued into all its deeper phases the question of using scientific thought for the solution of social problems, merely a pleading for quiet times. It is not outside the ironic possibilities of the human creature that one should get emotional over the question of moderating emotions. But that the destruction of many values and the enslavement of men have come from the immoderate practice of virtue is not an irony nor a point made for mere polemic reasons; it is a matter of observation by scientific method. The thinning out of loyalty into Royce's doctrine, loyalty to loyalty itself, does not go far enough. We can be loyal to our culture in its organic completeness and to all those parts of its diversity that we have been allowed to choose. But it is possible, because it has been done, to educate men beyond the ancient universal myth of the "chosen people" into an enlightenment of belonging not to the parish but to the world.

This enlightenment, as I said at the beginning, is subtle and difficult to explain as well as difficult, if not impossible now, to achieve. But we may begin to see some of the obstacles in the way and some partial success. Brotherhood, a feeling of warm and understanding intimacy with all manner of men and women and children, is evidently too exacting an ideal for men of any kind that we can now produce. Certainly neither Christian nor Buddhist education has ever produced more than a few persons who behaved by the rule of brotherhood. Without discussing how much the churches and the priests have sophisticated all primitive religious doctrine, we can say that it would be hard to find a culture in which the dominant, or even one among many recognized patterns, is to do unto others as you would have them do unto you. The doctrine of love is much too hard a doctrine to live by. But this is not to say that we have not made progress. It could be established, I think, that the next best thing to the rule of love is the rule of sportsmanship. We have gone a long way toward making that a common pattern of behavior in some cultures, and it is in those cultures that some degree of freedom is achieved.

Some perspicacious historian will some day write a study of the age-old correlation between freedom and sportsmanship. We may then see the importance of sportsmanship as a form of enlightenment. This virtue, without which democracy is impossible and freedom uncertain, has not been taken seriously enough in education, although the real success of modern education in training citizens for self-government has been training in living by the rules of the game. Montesquieu was probably wrong when he said that the characteristic virtue of a democracy was love of the public good. That is a virtue easily diverted to the public loss, if men are still obstinate in their ideas of what is public good and are avid enough for power. Freedom can be the condition of government only if a struggle for power, whenever it sets in, is regulated by mutually acknowledged rules and if, in addition to the rules, there is an honest willingness to abide by them and take defeat without violence.

A famous British physiologist, J. B. S. Haldane, once remarked in a conversation, and may also have said it in print, that the greatest contribution of the British people to political development was the idea of sportsmanship. It was a more important contribution than parliamentary government. In this, Haldane may have been enlarging the role played by the British, for certainly the Greeks had the idea of fair play as they also had a notion of democracy. And he may have been underestimating the importance of parliamentary institutions. But the truth remains that freedom can live only in the minds of men who feel a moral compulsion to behave in political and in all other contests as a well-trained athlete behaves in a game.

The virtue must be deliberately taught and deliberately learned. The institutional complex, the set of social habits of which cultural parochialism is one part and unfair behavior toward one's personal opponents is another, can probably not be all eradicated by precept or by condemnation. It may even be a part of the "original nature of man" in the sense that most cultures show it, and only those are free from it that have taken positive steps to train the young in other patterns. But such training is most likely to be successful if the pattern is made up of a series of separate but closely related habits, learned each for its own sake and established by reason of its own loyalties.

There are men who would not think of ending a baseball game that was lost according to the score by winning it in another way—for example, by bashing members of the other team over the head with the bat—but who would stop at nothing to win a political election and would falsify the records of voting if they got the chance. Such men may not be ordinary men. It is much more in accord with what we think we know of men that those who have a strong sense of decent behavior on the playing fields will have a sense of honor in politics.

The point for the teacher is, however, that we do not leave such things to the mere agglutination of the different traits of a personality. The educational system that has enlightenment as one of its purposes will teach sportsmanship on the playing fields of the young because politics has not yet entered the minds of children and it will not fail to point out that sportsmanship and decent civic behavior are the same basic virtue. It will go on in older years and even into all the circles of adult experience with counsel and example of the profound decencies that are expected of a free man.

If this appears to be giving to a boyish character an importance it does not deserve, it may be well to point out that if the sportsmanship of boyhood does not carry on into mature life, other boyish patterns will. The revulsion felt by most free men and women against certain phases of tyranny is partly a revulsion against the boyish pattern of the bully. The value of sportsmanship, as a virtue that can be taught to the young with some hope that it will be carried over into later and more dangerous years, is shown in international affairs. Empires would be less likely to exist and would in any case be less evil if all strong and imperial peoples had the sportsman's ideal. The fact that the British imperialists, with all their faults, with all their brutalities even, have still been more beneficent in real administration than other tyrants is a clear result of sportsman's virtues.

Aristocrats often have this virtue whatever they may lack. Time of war is a good test. The upper economic and cultural groups are often "international" in their sympathies. This is sometimes an outsider's comment. It may, even when it is hostile criticism, be a tribute to the lengthened ethical radius that persons sometimes attain when they have enjoyed power and travel and international, intercultural experience. In time of international conflict, especially in actual warfare, these beneficent attitudes are given up because of a natural desire to hold on to leadership in the home group (now hot for war), and effective internationalism is impossible. But these high-placed patriots may seem less interested than most citizens are in hate and atrocity stories. They are often suspected. "Treason in high places" is always common suspicion in war fever. We can hope that larger and larger proportions of each national group will learn some kind of intercultural sympathy so that finally demagogues will be powerless, even when there is provocation to war.

Many of the intellectual and sophisticated circles of all cultural groups give up a simple cultural provincialism and put in its place the provincialisms of caste, or fashion, the snobism of the "intelligentsia." These types of group exclusiveness are often ridiculous in themselves but they are actually less harmful to humanity as a whole than are the deep simplicities of a too narrow *ethos*. They are ephemeral and hence can claim no such sincere devotion. And sincere devotion to a narrow idea is the virtue that is here in question.

IV

Education, in our meaning of the term, goes on through life, providing knowledge and understanding and freedom, as the crises of life and society make new demands on the individual. Knowledge is power only if it can be used and only when it can be used. All kinds of knowledge are not equally powerful and any one kind of knowledge is powerful in different degrees at different levels of development. The members of a society may be helpless in confronting a danger, in spite of possessing the knowledge that would, if used, overcome it. This is often the result of cross-interests, as when fire regulations that we know will save lives are not enforced against greedy or indifferent landlords. But this is not the chief difficulty in making knowledge effective in our daily lives. The relation between knowledge and action is an ancient philosophical and educational puzzle; an old handicap in the building of cultures. In a society like our own, so largely urban, industrial, and democratic, the handicap is costly and serious. As we make our ways of living more complex, as we increase the size of our political units, and as we heap up more and more knowledge of the physical world, this oldest of questions becomes more dangerously insistent. How can what some of us know be most useful in the lives of all of us?

Three aspects of this question are now of evident importance. They are, first, the relation of socially held scientific knowledge to the management of each personal life; second, the relation of the knowledge of the few to the needs of the majority; and third, the relation of the knowledge of experts to the exercise of administrative power.

The first is chiefly a matter of pedagogy, technically considered. Teaching men and women, children and young people, what they need to know of their bodies, their minds, and the physical world about them is the standard task of schools. Much more than ever before we need to extend the learning period into mature years and we need to have institutions whereby knowledge can be gained whenever it is needed. The typical instance is the kind of knowledge that makes for what we have come to call "mental hygiene." The difficulty is not for specialists to learn what needs to be known; nor is it, in any great measure, a problem of any person's own knowledge. In this special relation of knowledge to action, the difficulty lies in each man's using what he knows. It would be a much more peaceful and progressive world if all men acted on their best knowledge of how to act.

We are involved here in problems of the will, of morals and ethical principles, and no doubt also in questions of religious attitudes outside our province. But if our educational tools and institutions are good enough to give the members of our group both courage and curiosity, and a strong loyalty to honest thinking, the solution will be possible and it will be in line with the slowly bettered ways of doing an age-old task, preparing the young to live.

The other two aspects of the problem are of more general concern and cannot be met so easily with institutions now in action. To try to give to every member of society the largest possible share in the knowledge held by the elite is a new enterprise. It grew out of humane impulses, economic needs, political pressures, and like complicated forces which have been at work in the Western world for about a century and a half and in the Eastern world for about a generation.

Mass education as we think of it was impossible, of course, before the industrial revolution. Men could not produce enough to support institutions of general enlightenment until they had learned to do work with natural energy applied to machinery. The schoolmasters of the ancient world were slaves because slavery was the only social and economic position into which a person devoted to teaching could be fitted, and society as a whole was not rich enough to provide schoolmasters for all its members. To provide the books, the equipment, the buildings, and the teachers for the simplest primary schooling for everybody demands riches far beyond the possibility of any culture before our own time. When it comes to maintaining institutions and tools of the scope that we need to keep our whole body of citizens aware of our socially possessed knowledge, we have to have not only great social wealth but great cultural ambition.

When they began to teach the mechanics how to read in the factories of Scotland and a little later in England, the owners wanted their laborers to be able to comprehend simple instructions and in their meager leisure hours to read the Bible.[6] The factory owners were afraid the workers might learn to write; that would lead to writing letters and letters can carry ideas and foment conspiracies. But knowledge tends to grow when once a beginning is made. In industrial communities now, the employers, trade unions, religious and philanthropic agencies and the government, all have a hand in helping adults to understand their world.

We have failed to live up fully to the promise of this movement. We have not learned enough about the interpretation of difficult subjects for the layman. We have not, with all our schooling, as yet produced many men and women who care enough about their own welfare to remake their institutions into better instruments of enlightenment.

This question of motivation is subtle. I am not overlooking the eager and often energetic zeal with which large numbers of men and women respond to the chance to learn something of which they can appreciate the value. But as we have already said, our democracy does not finally succeed in this ultimate and critical phase of our democratic purpose. If all forms of democracy are institutions devoted to the growth of personality, then the daily life of every citizen in a country like the United States

[6] J. W. Hudson, *History of Adult Education* (1851).

of America should be so involved in affairs both intimate and important that he would find it urgently necessary to keep on learning. But the centering of power for the sake of efficiency makes this difficult. Also those who have "public" purposes, partly because of the neurotic origin of the strong reforming impulse [7] and partly as the result of the standard mistake of seeking to create preconceived conditions instead of seeking to create conditions for the growth of persons, for these and other reasons, those who have "public" purposes are very likely to try to impose their innovations or regressions on the social group. Against these trends we have made little headway, if indeed we have ever honestly tried, and the members of our society get less and less natural stimulus to learning out of necessary public business. Sitting in study groups, listening to lectures, reading pamphlets, these are useful devices for those whose concern is active and who are looking for guidance in casting votes or discharging any other public responsibility. But there is not, in most of these enlightening pastimes, enough of motive, or responsibility in action, to inspire the sweaty struggle of real learning.

Every change in social development makes the drawing of the line between imposition and teaching more difficult, which is another way of saying that every change in the character or complexity of a culture makes freedom more difficult to determine because it changes the nature of the compulsions needed for stability. Both the reformers and the men of action, who have a naïve love of power, try to take advantage of this fact to encroach on the freedom of the individual, and the individual finds it harder to make his defenses because the more complex and the rapidly changing patterns are not easy to understand. The line between necessary compulsion and minimum liberty has to be drawn in terms of developing situations. The rule is to keep the line from getting fixed and taking on its own institutional importance.

The increase in scientific knowledge is in itself a complication

[7] H. K. Lasswell, *op. cit.*

in defining free action for the members of an efficiently organized society. Increases in scientific knowledge make more certain and explicit the consequences of choices offered by the plenitude of life to the members of a social group and so tend to lessen the number of choices the group as a whole will allow any individual member to make. This is exemplified in the control of public health. Before the certain consequences of contagion were known, neither the group as a whole, nor the legal government as its agent, had any moral reason for interfering with the behavior of a citizen sick with smallpox. After scientific information is available to the government, however, quarantine is a use of police power. We use restraint for the public good but all heretics have been burned at the stake for that same reason, the public good, to get rid of contagion.

The distinction between the suppression of contagious thinking and the suppression of contagious disease ought to be clear as soon as it is stated, but we are left uneasy. We do not believe in combating mere opinions with physical force. Good, but what are "mere" opinions? No thoughtful and scrupulous governor can always be sure what acts are dangerous, deserving forcible intervention, and what is dangerous opinion deserving only argument. The point at which an opinion becomes an act is not easily descried.

Will increase in scientific knowledge, then, by making prediction more certain make liberty more difficult? In any case, there is a free phase of this process, of this gradual restriction of the choices of members of an enlightened social group, which results from an increased knowledge of consequences. The possibility of freedom lies in the need to enlighten them as to those consequences and thus provide if possible that they act with consent. We might say that a person who has learned as a child that if he sticks his finger in the candle flame he will be burned has thus, by an increase in the knowledge possessed in general by his group as well as by himself, been restricted in his liberties. Society would interfere with anyone who tried to destroy himself by fire. We do not consider any such behavior sane. Society cannot give liberty to any but the sane, and sanity is definable as the condition of agreeing on the whole and in necessary essentials with one's fellows.

It quite well may be, in some future state of accumulated and dispersed knowledge, that a person who makes some of the choices now considered reasonable will be judged insane and locked up. This will happen, if and when it does, because our descendants will know so much better than we do what will result from our actions. But in that time, there will still be the free learning by the citizen of the scientific facts on which predictions of consequences are made, and he will in that period of learning be free in the sense that he will be learning more about himself and the world and the possibilities of their interaction. He will be growing as an individual, although growing into a pattern which is fixed at certain points in the mature phases of his development.

The techniques of interpretation for this diffusion of knowledge as widely as possible in the culture group have been slow to develop because they are nobody's business. The educators of this typical Western social group may be roughly divided into three kinds: the elementary and secondary teachers who pass on what they can find in the books; the "productive" scholars who seek new knowledge and train apprentices in methods of investigation; and the managers of institutions. Only by grace of temperament or accident is any teacher inspired to interpret what he knows to the public in general. Institutions for adult education,[8] of which there are increasing numbers, find their staff members with difficulty and often recruit them from outside the professional ranks. Interpretation of knowledge to the general populace, whether in adult education institutions systematically managed, or outside them, is mostly the concern of men and women whose methods are certainly not the pedagogical ways of professional teachers and are indeed mostly the hap-

[8] Lyman Bryson, *Adult Education* (New York, American Book, 1936).

hazard residue of earnest fumbling experience. Newspaper and magazine writers interpret physical science; reporters interpret political affairs; lecturers and priests interpret moral and social forces. The function of interpretation is served but it is more or less by accident and without any special selection or training of interpreters.

Culture would change more swiftly for the better if we could get men and women interested in those aspects of their own lives that could be enlightened by thoughtful and purposeful learning. And this would be advanced if some new social motives would induce scientific workers and scholars to take a pious interest in aiding instead of hindering the interpreters at their task. Often because they catch a frightening glimpse of what a hostile popular majority might do against them, or perhaps when they get a better notion of what real popular support for science might be if there were popular understanding, eminent scientific workers will read sermons to one another on the need to explain abroad what they are doing. But they would smile at the thought of devoting any first-class student to a career of interpretation. The business of interpretation is an ancillary and always slightly exasperating adjunct to the real business, to learn more and train bright young minds to go on learning more. The social judgment on this behavior, in a society where freedom is an ideal, must be toleration, but the social group will have its own reasons, if it hopes to create and keep its freedom, to get the work of interpretation done somehow. If scientific workers will not undertake it, others will be put to doing it for them.[9]

At the present time, when we know so little about the skills needed or about the psychology of free learning, it is quite impossible to say what this function really is. We can only describe it from the outside and on institutional levels. But we can at least dismiss the easy irresponsible dictum of the scientific workers who say that anyone who knows a subject well enough can explain it and the equally uncooperative but much more unpleasant

[9] Lyman Bryson, *The New Prometheus* (New York, Macmillan, 1941).

attitude of those who escape by saying that the whole enterprise of enlightening the ordinary mind with scientific information is hopeless.

Many phases of scientific application are impossible without the participation of everybody as, for example, the enforcement of the rules of public health. Sanitation is impossible unless practically every member of society knows enough to follow rules. Perhaps the rules would be more carefully obeyed if every citizen knew the reasons. But the real justification for the slow and costly effort to make the citizen understand his health as well as protect it is the democratic one. Informed action is better development for the person than the most scrupulous obedience.

The third thing we need to do to relate knowledge to action is to make knowledge effectively available to administrative power. This is an old problem and it is astonishing that civilization has produced no treatise on the special skill needed by the philosopher who is to advise a king. Plato saw no way of passing wisdom from the learner to the man of action except by making philosopher and king the same person, but his own experience in Sicily showed that this is not likely often to happen in the succession of tyrants and kings. Neither Plato with Dionysius, nor Aristotle with Alexander the Great, succeeded much in tempering violence with wisdom. Alexander may be admired for energy and military genius, which is like admiring a flood or a storm. These things may be magnificent; they are not wise.

Neither Plato nor Aristotle, out of their sad experience, tried to write a treatise on the methods by which the wise man can advise the king. And yet many men beside those two must have gathered personal lore if not rational skill in that dangerous business. When some savage chief asked his tribe's medicine man what would be likely to happen if he sent a maurauding gang over the hill to the next tribe, he was calling for something more than worldly knowledge; something more even than divination. He was calling for skill in interpreting wisdom for the uses of power. He was asking the shaman to predict the consequences of alternative choices in such terms as a man of action could understand.

It is true, obviously, that if a man could be both philosopher and man of action at the same time there would be no problem of relating science to policy. But more than constituent powers and special training are necessary to create such a being. Plato does not seem to have seen that men in power spend most of their time in holding their grip on that power; they have little time or energy for acquiring knowledge. There are examples of men who have had philosophic gifts and gifts for action also, men like Arthur James Balfour or Jan Smuts or Lord Haldane, in recent times. But while they were in positions of power they were far too busy keeping themselves there to foster in those periods their creative impulses in theory. No doubt, flashes of insight into the flux of action in which they were immersed were of great value to them afterward in making systems of thought, but even a Dionysius found when he was tyrant that he could never relax his watchfulness.

The press of routine business, which no man can altogether escape if he is actually exercising power, occupies the ruler's time and uses up his strength. Most of the actual exercise of his power is in passing judgment on the creative ideas of others instead of developing his own. This is as true of the men who control industrial empires, or vast mechanisms of mass communication such as newspapers and motion picture companies and radio chains, as it was of Dionysius.

Because they cannot take time off from watching their enemies and their competitors and because the actual exercise of power gives no leisure for thought as to how it should be used, the kings should not be counted on to be philosophers, and for these reasons, at least, they should not be expected to develop the techniques by which explanations of relevant wisdom are to be brought to them. This would be true even if it were conceivable, as it is not, that the temperament that puts men in power and keeps them there could often include a strong interest and a creative capacity for methodological invention. The methods of interpretation are the business of the man of knowledge or of some intermediary whose function can be clearly seen but whose qualifications are hard to list. We have indicated that we can be sure that the interpretation of science to men in general, the layman's version of the scientific truth, must be created by specially trained intermediaries since scientific workers are seldom by temperament, and practically never by training, capable of doing it. The interpretation of this same kind of knowledge and much that may be more recondite for the guidance of wielders of power is a task on a different level and possibly more difficult still.

The scientific approach would be to observe, and investigate, and devise generalized descriptions of, the ways in which advisers to rulers have succeeded in the past, such descriptions as can serve as a basis for the rules of skill. And I am thinking here of something quite specific, not mere wise lore. In facing any decision on a course of action, the ruler must be able to choose intelligently among alternatives and his adviser must be able to tell him the alternative consequences. A course of action is never a single act, followed by a net result and ending in an accounting for good and evil. A course of action is a curriculum leading on from one crossroads of choice to others in a long series; the decision at each branching of alternatives changes the whole prospect from that point on. Alternatives in action are not only dynamic; they are dependent in series, and very often the most important decision takes its importance not because of the immediate consequences but because of the alternatives which it will offer at some future stage of change.

It sometimes requires the deeper habits of thought of the philosopher to recall to the attention of the executive that, as each action is taken, the whole situation in which further action is to take place is more or less evidently changed. Any practical situation is like the bits of colored glass in a kaleidoscope; at a twist of position not one bit but all of them fall into new places and the pattern is completely new.

It is not the role of the man of knowledge, however, to tell the man of power what alternatives to choose. The king must make his own choice if he is a king. The work of the philosopher is to point down the various paths and tell, if he can, at each succeeding crossroads of the future what will happen as each path is taken, from alternative on to alternative, from choice to further choice. Those choices must be convincingly described; the man of knowledge is not a mere oracle who says, Do this and you will suffer thus and so. He gives the reasons for predicting, and translates his special knowledge into schemes of specially simplified principles that have meaning not only in the world of science but also in the world of action where the king lives. It has often been done, but no one has ever told us how, in general, it ought to be done to get the best uses of both knowledge and power.

In the educational phases of the growth of the person, the habits of freedom and the love of the institutions of freedom can be fixed. But a childhood education cannot last through life, except in a very simple and static culture.[10] For all lives lived in modern cultures new education is a recurrent need. And we mean education in the strictest sense, the deliberate use of institutions devised for the purpose of enabling men to benefit by the experience of others. In rapidly changing societies, the forms by which men live shift and change and men have to learn new ways, new occupations, new political ideas in order to live at all. Even if this were not so, the successive crises in each personal life, the crises of marriage and parenthood and age, make new learning necessary. No childhood education can meet the need because it is impossible for men to understand the remedies for problems if they have not yet met the problems in their own experience.

So the adult person lives in a network of influences (a flow

[10] Lyman Bryson, *Adult Education*, p. 5.

of symbols, we could say more technically) that are constantly modifying or confirming his behavior patterns, and some of these influences can be called education. For our purpose, it is not useful to draw the line precisely between adult education and any other influence, because the social engineer would make use of all influences. He would intervene in the flow of symbols whereever he could do so effectively and to good purpose. When a man reads a newspaper, is he educating himself or not? If we did have to decide we should probably do so by the criterion of the reader's purpose. But the social engineer would use the newspaper, as he would use motion pictures or books or broadcasting or any other medium by which symbols can be communicated, to intervene in the dynamic patterns of the loyalties of persons in his culture. He would try to persuade them to social change.

V

We can at this point make two *ad hoc* definitions to get us along in the argument. We can say that education is the formation of the habit patterns of the person and the fixing of loyalties to the habits that are shared in institutions. This definition is not adequate and may offend those who have not, in spite of my pleas, discarded the idea that habits are "automatic" or irrational, those who cannot accept the fact that personality is predictability. Second, we can define persuasion as intervention in the habit patterns to change their balance into a new configuration. One of the best examples in Western history of such intervention is the conversion of Roman citizens to membership in the fellowship of the primitive Christian church.[11] It is, like any other engineering operation, an intervention among natural processes in order to set some of them against others and produce a change. On the higher levels of abstraction it would be possible to say that keeping a pear from rotting by putting it in

[11] Donald Riddle, *The Martyrs* (University of Chicago Press, 1931).

an icebox, and keeping a man from doing something foolish by appealing to his "better nature," are the same operation.

Education fixes the potential energies of the human being in the patterns of behavior and emotional response. Persuasion modifies this dynamic pattern in daily living. The distinction is only partial but it will do. What then is persuasion?

Writers on persuasion, Aristotle to Bacon and down to yesterday, speak of persuasion as the process of moving the will, but it is hard to find, in what they have had to say, an operational definition of the "will" that will make possible a useful description of the process. For our purposes, an operational definition of the will may be attempted: We use the word "will" to designate the present balance of conflicting loyalties in the person. This means, of course, that the will changes whenever a different loyalty gets to be dominant in the person for any reason, or whenever the present dominance is appreciably lessened or increased. The operation by which we discover this balance is an observation of the person's trend in behavior. What he says about his motives is only a part of his behavior and, in many cases, not the most important part. We can discover conscious active conflict when he is slow to act; we can suspect, on hints from Freud, that there is hidden conflict in him if he acts with too much assertiveness. We see the perfect expression of will in actions that seem to be "instinctive," which is to say that they are acts that are not, as far as we can observe, preceded by deliberation.

This is not offered as a philosophical definition of the will and it certainly enters no candidate for the old circus of determining whether or not the will is "free." It is a trial at stating something that can be discussed in exact terms, an abstraction for scientific use.

Persuasion, then, would be the process of "moving," that is of changing, this balance of loyalties in a planned direction. Notoriously the man divided within himself is easily persuaded. The mind of the man who has no hesitation is a difficult citadel to breach. The steps in the process here analyzed are generally followed by successful intuition.

The beginning of persuasion is to learn enough about the whole pattern of loyalties of your subject, that is of his emotional pattern as related to his institutional and personal habits, to know which of his latent emotions are favorable to the action which you propose. If he is acting without deliberation, these favorable elements are in abeyance; he may even be unaware at the moment that they are in him. They may be either ethically lower or higher than the loyalties by which he is acting. For your own moral good you should appeal to the highest motives available, but there is no scientific reason for repeating what is uncritically stated in the rhetoric books, that it is "more persuasive" to appeal to motives that are higher than those on which a man appears to be about to act.

If such favorable elements can be discovered they must then be made evident also to the subject himself; that is, he must be made aware of the loyalties he is suppressing. The first reaction to any such reminder of motives that have been shoved aside is likely to be annoyance because what had seemed to the subject to be a simple choice has now been described to him as a complex one. His "will" was a moment ago quite firm, but now it is irritated by the hint of uncertainty. The gentleness, the tact, the lack of aggression and insistence, that have always been thought of as the manners of successful persuasion, are needed at this stage more than at any other time. An angry rebuff is proof that the conflict has been aroused in the mind of the subject but proof also that it is producing a discomfort that will be blamed on the advocate of the new demand.

The answer to such irritation, the next step, is to praise the loyalties that have been recalled from neglect. If the persuader is honest, this praise can be honest. He is urging the subject to do something he believes to be good and he can praise the emotional factors, the habitual responses in his subject, which are favorable to his proposal, as against those habits that are against his proposal. He builds them up if he can, if he is still being listened to, trying consistently and with every kind of favorable association that he can suggest, to build up these factors in the conflict in which they are now engaged in his subject's mind. Praise of them is his ammunition and he enters the subject's mind to fight on their side.

At this point come into play all those elements of persuasion that Aristotle talked about, the emotional proof and the ethical proof and the logical or reasonable proofs, which at any other stage of the process would probably never find a reception in the subject's mind. But persuasion can be quite valid even when it has no concurrent elements of logical proof. In any case, the ultimate usefulness of the logical proof is to be such an analysis of the situation as will show the subject what important factors in his own motivation are being overlooked or outraged by his course of action and to establish that these factors would, when reestablished in his present attention at their full value, be more potent than the motives that are now ruling him. If a flash of intuitive appreciation of a forgotten loyalty, springing out of a reminder by the persuasive friend or out of some accident of the mind, can "move the will," the result is the same. The action is the same; the motive is the same and has been more quickly reached.

Since persuasion is, on the part of the persuader, an effort to shift the pattern of his subject's loyalties in such a fashion as to make him see a wider range of relevant motives, it is generally a force working for rationality. The lack of rational conduct in so many men at so many junctures in their lives has been an affliction to teachers. Thinkers have too often made mistakes, and reformers have made blunders, by expecting men to be more rational than they are likely to be. These mistakes are the result of failing to allow for the inertia of the person as a whole, which often leads to what the observer can see as irrational action. The cause of this foolish behavior is that the subject has not shifted his attention from those of his own loyalties that have become irrelevant or unimportant to those that are now of most value to himself. There are many vocabularies that may be used to explain such actions as found in Pareto, Freud, or kindlier observers. The operational fact is simple enough for use; our will is often a pattern of motives which is not adjusted to the useful factors in the practical situation. Persuasion applied by leaders or teachers makes the adjustment. The will is moved. Rational action becomes the tendency of our desires.

Persuasion can work evil, of course. There is no safety for the complex mind except the best possible choice, and that is only the greatest possible safety, not safety complete or absolute. The social engineer has to use this tool at his peril and on his own moral responsibility. But even though he may be working to change cultural forms, to modify the behavior of persons in such ways as to change institutions and so make the new ways of behaving predictable, the social engineer is still carrying out the purposes that men fulfill in culture. If he is working for freedom, training men, in the educational phases of their experience, to demand and love freedom and especially to love those institutions that safeguard the freedom of others, and if he can intervene in their daily lives to persuade them actively to freedom, he is an agent of the culture. He is an active factor in the trend toward individual self-realization which is, in our naturalistic theory of freedom, the end of culture as the combination of all secular purposes.

PHILOSOPHIC BASIS

I

IN THIS BRIEF AND MEAGER INTRODUCTION TO SUCH A SUBJECT there must be great gaps in the argument. They cannot even be systematically placed so that a skeleton of what should be said can be perceived. This is particularly true of the philosophic roots of our generalizations. To find these it might be necessary to regress to first principles, and even in this essay the deepest principles cannot be entirely assumed.

What is mind, for example? How can one talk of freedom when he is using "patterns of habits" as the units of social life, and urges that we create new men for a new world in which we guess that we never could be happy? I cannot answer the question resolutely because too much work remains to be done. I can, however, discuss mind as something in the natural world that acts and so can be observed. We will make the attempt to place mind in the natural world because mind is the seat of freedom. It is in consciousness that one rehearses the symbols of possible choice, where one experiences pursuit of the end chosen and where one is aware of punishment or reward. By mind here, I obviously mean active consciousness and the possibility of active consciousness. This belongs in the natural world, the world of animals, plants, continuities, and atomic play. Whether it does or does not belong in other worlds depends, of course, upon what other worlds there are and what is meant by saying that other worlds exist. This, again, is not part of our subject. It is in the natural world that we are trying to make men free.

This philosophic view of mind has, however, a pragmatic purpose. It is a further and more difficult phase of the attempt to find for ourselves a clear enough concept of what we mean by freedom, to know well enough what we mean by our words, so that when we set out to achieve it as a value we may get as close as possible to our goal. We are attempting briefly to get at the roots of the concept of mind but to avoid hypostasis and myth. It is difficult. In most of our thinking we demonstrate that the natural working mind escapes into myth on the periphery of experience in any direction, the future or the past in time, or wherever our access to observation, direct or by report, is cut off. Not many minds, if indeed there have been any, are well trained in understanding themselves. How can minds adequately understand themselves since a mind is a duration of consciousness and is thus made up of the two things science cannot yet describe—duration and consciousness? These things can be experienced, not symbolically elaborated. But we can, nevertheless, observe dynamic phases of our minds; better still we can interpret mind by watching the behavior of other men and by studying the records of past action. The high levels of generality on which my own comments are made is an acknowledgment that there is no knowledge to which I have found access that gives concrete facts—entities that can be univocally described and quantitatively observed.

Consciousness in duration is a phase of human behavior. It is also, as Dewey says, a set of organized meanings. I would add that those meanings are the conscious concomitants of the habits and loyalties which may or may not be symbolically or otherwise present in the conscious thought. The use of this idea for our purpose is to give, if possible, a sense of naturalness to the working of the mind so that the place where the inward action of freedom happens, the locus of the experience of freedom, may be found on this earth.

Within the limits of possible response set by the innate, the physically inherited capacity of any person, experience is recorded in general ideas. This is a cumbrous but more exact way of saying that we learn. It is helpful also because it breaks up the process into two phases. We cannot learn unless we can record what happens to us; we must remember. But completest possible recall of everything that has affected our senses would not be learning unless we understood some of it. That is, we have to be able to see in the new experience some resemblance to the old, even though much of the new is different from anything ever sensed before. We perceive not only facts but similarities and differences among facts. We get, somehow, general notions, and live by them because they are an understanding, a mastery of life.

The recording process seems to be, for the present, beyond the reach of scientific examination. We can observe only the results; the apparatus is hidden. Anyone who tries to describe memory is a bold man indeed, and observers have been so endlessly astounded by the miracle of recording that we have given very little thought to the still more mysterious process, forgetting. It would be much simpler if we could remember either everything or nothing. The trick would be easier to understand if sensory experience were fresh soon after we got it and faded gradually, all details going out together.

What happens is much more complicated. We remember some things vividly and forget many. What we can recall at any moment may be different from what will be recalled at some other time. Any picture from our past is made up of all kinds of things, some fresh, some faded, and their freshness is by no means merely a question of their distance in time. Above all, we blessedly forget the greater part of everything that passes through consciousness; it would be intolerable if most of our daily experience did not go into swift oblivion.

Memory then is selective. Some day we may know how and why, but we must act now on the principle. And most of memory in a human mind is (mysteriously again) organized in systems of symbols. Whatever it is that "happens" in the brain when someone asks us a question, there is a cooperation there between past sense experience and meaningful symbols.

"What was the weather like yesterday?"

"It was hot."

The sensations of heat may be remembered but the sensations of heat and the words "it was hot" are quite different things. Philosophers are still wrestling with the question of the degree in which thought is dependent on language.[1] We go smoothly through our ordinary affairs without trying to think except by the use of language. It is practically impossible to define the symbol "symbol" but a child uses symbolic communication and organizes his experience in terms of symbols as naturally as he grows.

It used to be said that man was the "tool-using" animal. He does use more tools than any other animal and his gadgets are more complex. But a bird uses a tool when it builds its nest; a beaver when it builds a dam; a monkey when he throws a cocoanut. What is the nest of a trap-door spider but an elaborate and beautiful tool? A much more useful distinction between man and his brethren is that he uses symbols. So far as we know, no animal but man can make a noise that sounds like one thing when he means something quite different and still be understood by his fellows.

Of course animals do speak to one another by cries, and any uncultured alley kitten knows what I mean by "Scat!" The first point is difficult. It is possible that some animals do convey "meanings" by symbols. It is reported, and no one but a naturalist would dare question the report, that some birds post sentries while they feed, and the sentry's signal is not a direct call but an agreed upon sign. That may be true. But if true, it is so rarely true and its manifestations are so meager that they scarcely change the rule. Human beings also respond to tones. Human beings respond actually to both tone and symbolic meaning. Most, if not all other animals answer only to the tone, be it growl, mating call, or a master's voice.

In any case, man is the creature who constructs great systems of symbols called languages. One might even go so far as to say

[1] Wilbur Marshall Urban, *Language and Reality* (New York, Macmillan, 1939), p. 375: "Hegel said language was 'the actuality of culture.'"

that he builds great systems of symbols called cultures. If the systems or patterns of mental facts are organized by means of symbols, mostly language, then cultures themselves are systems of symbols. The meanings attached are understood in minds communicating with each other.

We can guess with some assurance at many ancient facts by looking at the skulls, the stone tools, and the burial hearths of our remotest ancestors. If we knew something about their speech we would not need to guess so much and would know much more. History began not when men began to write history but when they began to write. They began to record their symbols. Their speech was no longer lost as they uttered it and the uses of their tools was recorded to be divined by those who came long afterward. More important, their purposes were recorded for us to understand. In the same way, we can know by language those distant from us in space as well as those far back in time.

As far as can be seen, all animals but men live out their lives in one environment. They never know anything, outside of instinctive responses, except what they learn either by direct experience or perhaps by some exemplary action of their parents. This parental teaching is very meager and can be done only by demonstration.

Man, however, lives in three environments. He has all the opportunities for direct experience that are open to any living thing. He can try out his inventive ideas and make good resolutions when he gets his fingers burned. He can watch his elders and his betters do things that he would like to do, and imitate them, as soon as he has learned to imitate, inside the limits of his powers. He has an advantage, even in making use of this first environment, because he has more responsive nerves.

Also, beyond the reach of anything any other animal can compass, he has the power to live in the experience of all of humanity gone before him, in so far as it is symbolically set down. This is his second environment. He need not discover all over again what Euclid knew about the structure of "space" nor what Plato knew about the good life. He does not inherit the past, as does the ant, in the structure of his nerves. Experience detonates the behavior of the ant instead of teaching him. Man has to learn all he knows but he learns mostly the residues of other's learning and so experience accumulates. The fact that he is not instinctively adapted to the subtle difficulties of his present environment, but must learn and can learn not only of this environment but of others also, is the measure of what we called his advantage of being modifiable.

In space also, he can stretch beyond himself and by his symbols learn from his third environment. Other animals can learn only from what strikes their senses and only the direct meanings of those impacts. But I can read in a newspaper of something that happened a thousand miles away and learn, in that third environment, truths that may never come by any solid evidence within my senses' range.

More than any other animal, man changes his environment in the larger general sense by living in it. His culture is made up of his organized and conscious ways of controlling it and changing it in order to control it more definitely. Of course, men and cultures often fail, and the general environment often defeats them, but it is changed in the battle. From the standpoint of social management, however, the most important point to note here is that man can only affect his surroundings—animal, vegetable, mineral—in collective action, in collectivities held together by symbolic communications.

With symbols, then, man in culture thinks and gathers a loose collection of meanings in which some conformations take on harder shape. Has man always thought as he tries bravely now to think? Was there, as Levy-Bruhl and others have believed, a "pre-logical" phase? We can doubt it. Here we can only speculate unless we want to call some of the primitives of today prelogical, which would mean only that we are baffled by their logic. It is important to speculate, however, because we want a clear notion of the relation between the simplest kinds of thinking and the origins of institutions.

There is a certain amount of danger, of course, in trying to find the character of our own ancestors by looking at simple folk who now live here on earth with us. "Contemporary ancestors" is a sharp but misleading epithet. There appears to be no decisive reason for thinking that any culture now existent is any "younger" or less "mature" than our own. We have reached what we now have and now are (so far as we know) in the same number of years as it has taken some of our brothers to learn how to go through a few dance steps and use a breach clout. To be sure, some naked barbarian such as the Australian black may have a spiritual lore that seems more like the achievement of ages than does his meager handful of gadgets. But even his mythologies seem to us strange blends of insight and obscenity. It is hard to believe that he has come as long a road as we have but the anthropologists tell us that he has probably spent as much time on the way. Hence, although we can take these contemporaries as examples of humanity and pick their cultures apart since they are simple, we cannot be sure that we are descended from any beings like them in patterns of living. The very fact that we are different now indicates that our own primitive phases may have been different then. The essential difference between our primitive phase and their culture may lie exactly in that point, that we have gone on into complication and development. And, in any case, we can learn something about the way our own ancestors thought by looking carefully at how we ourselves think, trying hard to avoid the illusion that we work, when untrained, by any rigorous method.

Logicians and psychologists have given us analyses of what we call "thinking" and the result has been mostly the complication of a natural activity. I would not offer one more of these descriptions if it were not for the fact that we need a defined concept to work with and especially because I want to make clear the hypothesis that the thinking of the past, all the past, no matter how primitive, has been essentially the same operation. This will help us to put the working mind in its natural place.

Without stopping here to examine the nature of perception itself, that is without examining the data of immediate experience, we begin by considering memory because thinking always involves at least two situations. One situation confronting us when we think is, as Dewey says, puzzling. The other situations, all *remembered*, are examined as we try to find in them some similarities to the puzzle we are trying to understand.

But we do not remember things in clear strands of equal brightness or in mosaics whose parts have all the same definite outline. We recall the past in loose aggregates.[2] Our memory, whatever its mechanism, re-presents, to consciousness, loose aggregates of the perceptions, the experiences, of the past. The process is immensely complex, of course. It cannot be made simple. But we can with gross accuracy, say that we re-call loose aggregations of the perceptions, the experiences of the past, and that they are brought back with sharp centers of interest which are adjusted to our purpose in trying to remember. That is to say, we might recall the same event at one time and have a center of interest at one place in that event, and at another time remember the same situation with quite a different center. Wherever it happens to be, the center is likely to be related to the meaning of the experience when we had it as well as to our purpose in trying to remember.

When we "think," we are facing a puzzling, an unresolved present aggregate, but are never quite sure what elements in it are of most significance. That is, we never know exactly and completely what elements in it are most relevant to the answers we seek. We are always like a lumberjack searching in a log jam for one log that will, if moved, let loose the mass. We never know completely nor exactly where to find that key fact or set of facts when we start thinking. If we did we would not have a problem. But we know more about some situations than about

[2] Peace to Pareto, I am using the word in its ordinary dictionary sense!

others. And the quality of a situation that makes our search for the key facts easier is that it contains a large number of elements which we recognize as the same, or effectively similar to, some important elements in previous situations for which we did find an answer.

Confronted with a puzzle, what do we do? Try first to define exactly the nature of the difficulty, Dewey says. But this, I think, can describe only the working of a mind of considerable training and explicit skill. Our naïve process of thought could probably better be described as trying to remember a similar situation, to find some part of this new puzzle that bears effective resemblance to a mastered puzzle of the past. In fact, there need be no clear idea of the puzzle itself but only discomfort. And in searching our memories for help we may not consciously be seeking a puzzle situation in our history but something vaguer and more emotional, a feeling of triumph that we remember having had and the events that went with that feeling. Here is the root idea of cause. An idea of cause and effect, of necessary or invariable concomitance, is always what we want, of course, as the key for our present problem. But what are cause and effect, naïvely considered, but members of the same remembered aggregate?

Let me illustrate from what might well have been the experience of one of our forefathers, before Aristotle but not necessarily before logic. I am going on the hypothesis that he thought as we do.[8]

This ancestor of ours went out one morning and saw from his cave door the slow mist rising from the Dordogne. There was a

[8] I am quite aware of the fact that modern ethnologists (e.g. Radin, *Primitive Religion*) use an abusive word "psychologizing" to describe any attempt to reconstruct the thinking of a primitive ancestor in the hope of getting at something basic or essential in mental habits. My defense for committing this crime is that (a) I am not guessing at the behavior of modern primitives instead of observing them. I am talking about ancestors that were chronologically as well as culturally our forebears and out of reach; and (b) that this is in any case a fable by which I hope to show how we ourselves think and I can rely on candid self-examination to provide the necessary criticism.

cloud in the sky and its shape, he noted, was much like the shape of a bear. His woman spoke to him; he noted her words and her tone. He bore his favorite hunting weapons and he had pronounced the proper incantations. He went forth to hunt. When he came back in the afternoon he brought meat.

If you and I, just as we are today, had undergone a similar experience we would afterward "think" about it just as he did. If there was nothing unusual in the end-result, that is, if we had no more and no less than usual prosperity in the hunt, we would not think about it at all. But if it ended in surprise, we would remember it differently and it would differently affect our future.

Let us, out of pious kindness to his imagined shade, suppose that the old hunter won a mighty victory. Suppose he came home with a bigger kill than ever before, nutritious, dripping red, delightful!

After a few days of repletion, after the sleep and the slothful pride, there is nothing left but bones. His wife speaks to him again. He is, we suppose, an experienced hunter. But never in his experience had he ever hunted so well. And as he starts out again, he *thinks.*

There is even a puzzle of the Dewey kind hidden here although it is vague in the hunter's head. He is looking for the answer to the questions that we, logicians that we are, would phrase as: What was the *cause* of that great success last time? How can I make it happen again? And he asks himself what is really the same question when he says: What can I remember of the way things happened that morning so I can make them *all* happen again?

I believe that the most naïve and primitive minds work without question on the principle of a uniformity in nature. They remember things in aggregates and take for granted that things come so in the world, tied in bundles of events and circumstances. If you can reproduce some of them, you can have the others happen also. So the thinking of the hunter was to remember everything he could, and his entirely logical procedure was to repeat the events of the happy hunting day as far as was in his power.

We are often told by modern psychologists that thinking and doing should not be separated; they are two sides of the same organic action. It seems very likely that the primitive man we are now imagining would never have thought of separating them. He probably tried to do whatever he could think of doing, as fast as his memory supplied him with pictures of the past, of that triumphant past situation which he was trying to reenact. And if he took it for granted that the repetition of some elements of the past situation would have the effect of recreating the whole aggregate, with its bright center of happy hunting, he was thinking with as much "logic" as any man has ever been able to show. The difference between his thinking and that of a physicist working in a modern laboratory is not in logic but in knowledge. At this point, I believe, Dewey is not only unjust to the primitive but not quite precise in his description of the thinking process in a practical situation.[4] He distinguishes between knowledge and habit:

> Knowledge is a perception of those connections of an object which determine its applicability in a given situation. To take an extreme example; savages react to a flaming comet as they are accustomed to react to other events which threaten the security of their life. Since they try to frighten wild animals or their enemies by shrieks, beatings of gongs, brandishing of weapons, etc., they use the same methods to scare away the comet. To us the method is plainly absurd—so absurd that we fail to note that savages are simply falling back upon habit in a way which exhibits its limitations. The only reason we do not act in some analogous fashion is because we do not take the comet as an isolated, disconnected event, but apprehend it in its connections with other events. We place it, as we say, in the astronomical system. We respond to its *connections* and not simply to the immediate occurrence.

[4] John Dewey, *Democracy and Education* (New York, Macmillan, 1924), p. 396. Reprinted by permission of the publisher.

There is, of course, no reason for disputing this point merely to get justice for the savage. It is too late now for us to rescue his repute among his physical and intellectual descendants. My point is rather that there is in this passage from Dewey what seems to be a confusion regarding the principle basic to all thinking, primitive or modern.

The savages do not merely fall "back upon habit." They think in logical and reasonable fashion within the limits of their knowledge. They are faced with a puzzling situation. In so far as they think (and we should not credit them with more capacity for calmness under strain than we ourselves have now) they will examine the situation. In another place Dewey himself suggests the answer,

> The next step is suggestion of some way out—the formation of some tentative plan or project, the entertaining of some thing that will account for the peculiarities in question, the consideration of some solution for the problem. The data at hand cannot supply the solution; they can only suggest it. What, then, are the sources of the suggestion? Clearly, past experience and a fund of relevant knowledge at one's command.[5]

The real difference then between the drum beater and the astronomer in the presence of the comet is in the difference between their past experiences! The savage is not the victim of habit but of ignorance; he has no sufficient "relevant knowledge," and he thinks as well as could be expected with what he has. The astronomer is the possessor of specialized knowledge; he has been living in a culture that provides more and better knowledge relevant to the moment. It is not only the personal past of the astronomer that is superior but the whole cultural past of his people.

The word "relevant"[6] is most important because the capacity

[5] John Dewey, *How We Think* (Boston, Heath, 1933), p. 15.

[6] Cf. the discussion of "relevance" in logic in F. C. S. Schiller *Logic for Use* (London, Bell, 1929).

to think is a mathematical function of relevant knowledge. The savage has very crude ideas of what is relevant and what is not. He recalls loose aggregates of previous experience and tries to reproduce the missing parts that he desires, such as security from danger, by repeating parts that are in his power, such as making a great noise. The recalled aggregate is "danger–from wild beast–terror–noise–safety–triumph." Since the comet is entirely new he identifies the present with the past by the elements that are common, adds to them those that he can produce by his own activity, and expects happy results.

The astronomer does the same. To disregard terror and danger, to see that the comet is not a creature and therefore not likely to be scared away, to find relevance instead in "an astronomical system," are all the result of knowledge, properly used, of course, but still knowledge, not power of any other kind. How could the savage fit a comet into an astronomical system? He does not have such a system. If he has any astronomy at all it is still half mythology, and if he never saw a comet before he has no place for it.

The importance of all this is evident when we think of the factors in social change. Men born in different cultures have different powers, not because, as far as the evidence goes, they are different kinds of beings. All that is shown is that men who use what they know in about the same ways will have different powers in different cultures because different knowledge is available. The continuities and increases in those powers rest upon the effectiveness of the growth in culture, not on anything else.

II

Psychology is a young science. It will not go very far by speculation. It is now beginning to make its progress as all other sciences have done, by the slow piling up of facts whose validity and relations are subject to vigilant criticism. What is said here about the patterns of our thought is no contribution to science but rather the kind of empiric judgment that has to be made in any engineering operation in order that the business of the world can get done as science grows. Such judgments are easily discarded when knowledge is gained.

Man thinks, we said, in loosely remembered aggregates, and he makes patterns out of the material in his head, out of the residues of his impressions. These patterns are a natural expression of what appears to be his natural faith in the basic regularity in nature. He believes, or very early learns to believe, that items found together once will be found together again, and that if they come once in a serial order the beginnings of that order, newly perceived, can safely be taken as a signal of the coming of the rest. Nothing in Hume's criticism of the causal sequence changes, even for Hume, the "habit of causal expectation." Indeed, Hume thought that the expectation was a habit acquired gradually as other habits are gotten, although no one recollects the process.[7]

The faith in the uniformity of nature, the basic element in anything that can be called thinking, is stronger, we might say, in the modern scientific worker whose religion is likely to be a naïve pantheism than it is in many a theologian who cannot see his God as immanent in God's work and therefore allows for miracles. The working scientist has no doubt of the sequences.

We go much further in our normal thinking, however, than merely to record and recollect the loose aggregations of events. We grow up in a culture, which is another way of saying that we learn to imitate and take on the overt folkways that we see about us.[8] And also we learn to associate with these folkways the symbols and the symbolic systems that are fragmentary at first but mastered in maturation, abstracted and built up, until we can communicate with others in those subtle and marvelous structures that we call philosophy, science, and art.

The relations among these high abstractions and the tangible

[7] David Hume, *Works*, Vol. IV: *Inquiry Concerning Human Understanding* (1826), p. 87.

[8] Cf. N. E. Miller and John Dollard, *Social Learning and Imitation* (New Haven, Conn., Yale Univ. Press, 1941).

objects in experience are a matter of basic importance to the social engineer. For the moment, we need to note that experience is brought under control by the symbols that generalize, organize, and communicate the matter that is deposited by sense experience in the mind. Soon after birth the mind begins to organize its still meager accumulations so that new matter falls often into patterns ready to assimilate it. Sense impressions are generalized as soon as felt. By this we mean nothing more mysterious than our power to make a sum out of a prickly sensation on the skin, a drop of moisture on the brow, a slight oppression in breathing, a glance at a column of expanded mercury in a glass rule, a sense of discomfort that has no location—"Getting hot, isn't it?" It is the character of the mind to make patterns.

Whether or not the patterns that we make out of experience are "existent" in some realm of reality outside the nervous system is not part of our problem. The operational truth is that the mind makes patterns according to its own natural form as a mind. The surprising element in such a simple conclusion is that it took so long in the history of thinking about thinking for anyone to arrive there. The "gestalt" psychologists, whose technical conclusions seem to me not yet ready for practical use, are busy with analysis now, but it took a long time for us to conclude that, if all natural nonmental operations had natural patterns, then the mind's action, called thinking, would have natural patterns also. The difficulty seems to have been that we first took culturally derived notions as inherent, and as that naïveté has become impossible, we now take cultural patterns as unconditioned. What we are saying here is that culture and particular experience supply the content but that this nutriment grows into forms that are predetermined; they are characteristic of the mind in the same way that possession of two legs is characteristic of the human body. It might have been surmised at any time, not by analogy but by that same faith in uniform processes in reality, that the mental action of the nervous system would be patterned like all other organic behavior. The mind has a morphology and we are just beginning to see the shape of it.

Every live seed that can get food and air and its necessary conditions will grow. The form into which it grows is not a matter of luck, however; it has a previously determined life career as onion, cow, beetle, man, whale, oat stalk or trout. It may be killed or stunted but its normal "pattern" is all there. In the same way, not analogically I would insist, but as another example of the same process, the mind takes food (experience) and responds to the favorable conditions of growth by transforming its "digested" nutriment into previously determined patterns. What confuses this fact in our perception is not only our own participation in it but still more the greater freedom of the mind's organic mental powers, or the lesser degree of predetermination in its fulfillment of the possible patterns. We may quite possibly discover that this greater freedom is only apparent and that what really happens is that the mental organism is only more responsive than "physical" organisms are to differences in alimentation. The physical living body will make much the same kinds of cells and tissues and organs out of its food no matter what the food may be, as long as it is digestible. The mental facts are more variable. The patterns, in the most schematic sense, are organically predetermined. The content is determined by the experience of the person in the culture.

Whether or not the foregoing is in any sense true as an assertion of reality, an existential judgment, the practical judgment is that all mental contents can be observed to fall within discernible patterns and that human beings will behave as if this organic growth into patterns was real. That is the only kind of reality with which we are here concerned.

By observation, and subject to further observation, we may say that there are three basic kinds of patterns by which experience is organized in the mental life of the normal person. They are never completely separable in any one person's thinking; they are mixed as are the basic attitudes of adjustment and mastery.

Auguste Comte, with his three stages of knowledge, religious, metaphysical, positive, may have had a clue to this same idea but he seems to have failed to discover that what he took to be sequential and partly cumulative in the development of sciences and other culture traits was actually coexistent as a natural schematism in all the minds we know anything about. The three basic patterns can be called practical, absolute, and scientific. These, like all the others, are arbitrary definitions and the words used must be wiped clean of previously attached meanings.

Practical patterns in the mental content are arrangements of experience by which persons try to control events.

Absolute patterns in the mental content are arrangements of experience that determine all preferences.

Scientific patterns in the mental content are arrangements of experience by which persons try to predict events.

All patterns in the consciousness of a living being are mixtures and combinations, of course, and even if they were in purer form the patterns would not fit very well into older categories. The second class, the absolutes, includes the true, the beautiful and the good—all values. But in the third class, the scientific, there is only one value, the true, and truth is taken there in a quite different meaning. Thinkers throughout past history have used all three terms, practical, absolute, and scientific, in many different ways.

Any arbitrary definition can be explained. The practical pattern is the arrangement of his remembered aggregate that is made by the ancient hunter when he gets out his weapons and his rituals and sets out again to do some more good hunting.

The chief difficulty in making a clear extrication of this pattern for theoretical uses is that primitive behavior, in our ancestors or in ourselves today, is such an amalgam of means. Our three patterns are described by the principle of "purpose." They are distinguished from one another by the purpose in the mind of the person when organizing the residues of his experience into each one and when behaving in accord with that pattern's indications. They are patterns of attempts to control, attempts to choose, and attempts to foretell.

When a person is trying to control, however, he uses all sorts of means that are provided by his culture. He also rejects possible means for other than practical reasons. The Eskimos who are reported to prefer starvation to the eating of seal out of season are permitting an absolute judgment to override a practical judgment. The blasphemer who used a crucifix as a weapon would be acting in reverse fashion. But all of us, ancient hunter or modern businessman, scientist or poet, we all try to control events by skill, whether exerted on other persons or on things, and by compulsive rituals. These rituals may have religious trappings, they may be extracted from the institutional complex we call religion, but they are not used in the religious attitude of propitiation or bargaining by sacrifice. They are used not as means of adjustment but to force events to our will.

On the other hand, the true scientific pattern is a help and is not transformed when it is mixed in the practical pattern. The more operational knowledge a man has, in the sense of power to predict the behavior of events, the more power he has to control them. In fact, he could have no practical power whatever if he had not succeeded in making valid scientific patterns. But it must be clearly remembered when we acknowledge this dependence of the practical on the scientific that we often feel a dangerous temptation to reverse this mixture and modify the scientific pattern by the practical judgment. This is necessarily a corruption of the scientific pattern since as Cohen and Nagel put it, ". . . the *method* of science is more stable, and more important to men of science, than any particular result achieved by its means."[9]

A practical pattern is exactly the organization of experience which is formed for the purpose of achieving "particular results." The practical must be subordinate to the scientific or the practical itself will fail.

[9] M. R. Cohen and E. Nagel, *An Introduction to Logic and the Scientific Method* (New York, Harcourt, Brace, 1934), p. 395.

In one phase of this intricate problem we might usefully take a hint from Comte's three stages. He appears to have thought that religious, metaphysical, and positive were sequential in time, although each succeeding phase included something of the preceding phases. The patterns we speak of are not phases, either sequential or in process. They are simultaneous characters of the mind's action. But it is still true that earlier phases of recorded culture have shown a gradual purification of the scientific pattern, a slow process of extricating that pattern from the other two. And this, again, is important as operational truth because people of meager education find it difficult usually to make scientific judgments on unfamiliar matters and they are seldom capable of examining their own absolutes.

In a very real sense the progress of the scientific activity of men in culture, the growth of those institutions that are based on an attempt to describe events without adjectives of value, has been achieved by the gradual extrication of man himself from the event he is trying to control. The "objective" judgment is a deep and severe reversal of the naïve impulse to master brute fact.

The modern cycle is to make a scientific pattern, ideally free of either the absolute or the practical, thereby establishing an expectancy in events. With this operational knowledge, a person proceeds to seek an end that is indicated by an absolute pattern of values and he uses the means which his practical judgments indicate.

Primitive thinking reaches no such efficient analysis. In the most primitive thinking that we know anything about the means used to control events are not based on "objective" research at all. They are the results of remembered experience, of course, and they are culturally determined. But they are actually, in large measure, attempts to inject the person into the event itself in imitation of what is desired. The early pattern of the dance is practical since it is a form of magical compulsion, and absolute in the fact of its deep emotional velocity, and scarcely scientific at all except in the clinging fragments of operational experience. The sky is hard and the ground is parched and we need rain in the pueblo. We dance it; we dance the rain. And we dance the growing corn up out of the earth. And we dance the successful hunt. The dancing is a direct, although highly ritualized presentation of the event desired, giving shape to the event, not compelling nature merely by example. We speak of this as "sympathetic magic" using Frazer's term. It is clearly something much more. Symbolic and direct experience are confused. But it is an injection of the self into the texture of reality, into the very stream of events, as a way of compelling reality to give results desired.

One way of describing the progress of man's mastery of material fact is to say this: He has slowly learned not to inject himself compulsively into events by ritual, by dance, by emotional anxiety. Finally he is learning not even to inject himself into his own thinking even by "purpose." He is learning to achieve scientific patterns that have no intermixture of practical purpose and, being free of intermixture also with the absolutes, can give him descriptions without adjectives of value that are therefore dependable for prediction. On these descriptions and predictions he can base practical judgments that will work. And if his absolute judgments, his values, are too much out of harmony with what he has learned by his search after facts, he will be defeated in the world of common experience. What transcendental rewards he may have is not a matter of observation in the public world. The philosophical relations among these kinds of patterns are intricate and subtle and they are outside my purpose.

It is evident that this slow process, the extrication of the scientific type of thinking from the others, and its growing cultural importance, raise very difficult problems as to values. Scientific knowledge is recorded and organized experience. The absolute patterns of values are also recorded and organized experience, but science makes patterns that exist in the public world; absolutes, by their nature, are peculiar to the person.

Scientific patterns are corrective of absolutes whenever the latter disregard observable facts in practical action. Hence some modern thinkers, Dewey [10] for example, believe that we shall ultimately find the basis of all values in experience, scientifically organized.

There can be no doubt that this is true in the negative sense, at least. To illustrate, the National Socialists of Germany had a doctrine as to the physiology of race which they held as an absolute. It was a value, a basis of compelling preference. By its sanction they killed the Jews. Out of this, no doubt, they got satisfaction. All scientific knowledge could do to correct their judgment would be to provide the demonstrable truth that killing Jews greatly lessened the number of family strains of high capacity in the German nation. Jews were killed to "purify" the German national stock, but the result was and will continue to be loss of intellectual power in the German stock and no "purification" in any scientifically recognizable sense.

The Germans could, in defense, say that they did not want Jewish brains. That is the kind of value judgment that is free of all experience or evidence in the public world. But if they should make a series of practical judgments indicated by this preference and based on mistakes in the physiology of race, they would eventually fail in their practical purpose. This is precisely the same kind of behavior as is shown by a tribe of primitives who will starve rather than eat healthy food when it is tabu. But whether the proof that objective, operational truths can correct value judgments proves also that value judgments are to be found in pragmatic experience is another question, of much greater difficulty.

But we can note this paradox. The absolute judgments that are judgments as to Reality, or Beauty, or the Good, and which by their nature claim to be universal are found by observation to be relative to their cultural location. That is, the judgments on absolute Truth of a Hinayana Buddhist, born in Chieng-mai and

[10] John Dewey, *The Quest for Certainty* (New York, Minton, Balch, 1929), p. 33.

educated by the yellow-robed mendicants of his village temple, are Hinayana Buddhist. A grocer's son in Nebraska may claim similar, or competing, universality for the Presbyterian judgments of his family and his cultural location, and they will be Presbyterian. But something discovered in a laboratory in New York and stated as an operational, tentative "truth," will be true in precisely the same degree when tested in Chieng-mai or in Nebraska or anywhere else. The absolute is, in this sense, culturally determined, which makes it relative. The operational judgment is subject always to correction but not to any cultural difference. In this sense, the universal becomes relative and the operational the "universal."

THE GOOD SOCIETY

I

WHAT WE HAVE SAID IN THIS SOMEWHAT SPECULATIVE DISCUSSION is that the mind, when looked at with naturalistic candor, appears to be an organ by which the human being uses his experience in this fashion: Past events are remembered as loose aggregates that become better organized if we have some powerful meaning or purpose to which they can be related as of greater or less importance, and clearer if we have past knowledge by comparison with which we discern the elements of the aggregate that are relevant to the purpose. But the mind has a shape that is natural to it and determined by its nature, and in growing in this shape it will use the materials of experience to make patterns that guide purpose by determining choice, the absolute; that enable us effectively to describe our universe, and so to predict events, the scientific; and patterns that recall for our use the relevant elements of the past that will, when repeated, get us what we want in the future, the practical.

If this is one of the many valid ways of describing the mind, if it describes the mind for our purpose, it implies that freedom can be a reality if scientific means are used to understand the world well enough to foretell the consequences of our acts. Scientific powers and practical powers can collaborate in the realization of the conditions that make possible the realization of the values that the absolutes declare. It is true that either practical action (what we have been calling engineering), or scientific prediction of sequences, may bring an absolute into question because we wistfully seek realization of our absolutes in the day-by-day world. But in so far as this is true the fault lies in the natural universe and not in the use of knowledge and power for ideal ends.

It is the argument of our scientific humanism that the collaboration of the practical and the scientific in creating the possibility of the realization of values would make a society that was successful in realizing such values. Freedom was the absolute postulated for our discussion. More knowledge, stated in univocal terms and developed into quantitative generalizations about abstract entities, would make practical measures for freedom more likely to succeed, and there is nothing in the nature of the mind that would make this freedom an illusion. It would still be what freedom always has to be to exist in the real experience of persons in a real world. It would be rehearsal in the mind, by symbolic patterns, of possible courses of action, with substantial knowledge of possible choices and probable consequences, and with a reasonable hope of reaching the end finally chosen. No achievements can be guaranteed in this world by any system, no complete knowledge of either choices or consequences. Scientific humanism would as a philosophy, as a rule of action, encourage the greatest attainment possible.

This would make for a good society. What can we call a good society? Must it have freedom as one of the elements in its complex of values? In a free society there would be institutional patterns by which men could live free lives. The personal patterns of most of its members would make them want free lives. This preference for freedom and devotion to free folkways would be characteristic not only of the normal majority but also of a very large proportion of the deviate types in such a society, because part of its freedom would be seen in its accommodation of difference.

No doubt other kinds of social structures lacking freedom might be good, nevertheless. The whole persuasiveness of Plato's genius went into trying to make men want a society in which there would be physical and moral regimentation. He was pleasant in the *Republic* and petulant in the *Laws*, but he did not change in his desire to rid his ideal polity of freedom. He did not trust freedom. Perhaps Socrates did; we cannot be sure. But Plato's ideal republic had in it, nevertheless, great and eternally desirable human goods. If it were important to determine such a question we might, by our own analysis, be compelled to say that good societies were not necessarily free societies. We are saying only that a free society is good and have been discussing how to procure it.

There should be, however, some kind of calculus of goodness in culture which would be unaffected by any subjective or patriotic attachment to our own experience. If we should get to be really successful in social engineering, in achieving intentional social change, we should need a calculus. We should look for an objective criterion by which all cultures, our own included, could be freely judged. In such judgments we are now far short of the objectivity by which we measure less important issues. Trotter shrewdly remarks that in important matters we prefer certainty to knowledge [1] and our estimate of our own society is important. The best we can say is that there has been a vague reaching out after an ecumenical criterion, and it may be possible in some future day to construct one.

The basis for a calculus of the achieved degree of goodness in a society or a cultural group will be, I believe, this principle: Societies are good in the degree to which they make possible the attainment of the ends which their citizens are taught to pursue. The educational agencies, and all the other more or less self-conscious institutions of a culture, inculcate ideals, and all the institutions of the culture provide opportunities. If the opportunities make fulfillment of the inculcated ideals possible in high degree and for a large proportion of the group, there will be as much good in the lives of the members as the group circumstances can provide. If aims are taught that are impossible of

[1] W. Trotter, *Instinct of the Herd in Peace and War* (London, Unwin, 1917), p. 35.

fulfillment in that time and place the result is frustration and the society is not a success.

This criterion, which I have offered before,[2] has been roughly criticized as arguing that we should surrender all "values." That is not true, of course, but it is true that I am doubting and offering to deny that there is any one absolutely best social design. Value patterns in the normal working of the mind are, as we have said, absolute by nature. But we are dealing now with scientific judgments; they cannot be absolute. Observation of the various devices by which men and women and children have made their precarious way through this world, in all sorts of places and times, gives one the conviction that human problems can be solved in many different fashions, not equally good certainly, but not ranged in any certain hierarchy of better or worse.

The criterion does not lead to the doctrine that success is good for its own sake no matter what purposes are followed. It hangs on the belief that values that can be realized in actual living are good, and that evil values are self-defeating in their nature. Their certain self-defeat is the essence of their worthlessness. This is high doctrine, and difficult, and not congenial to the modern mind which has gone so far from Stoicism. It cannot be argued well in this brief space. It must meet squarely the question raised by actions that may be good for me and evil for you; profit to one at the expense of another. Moralists have tried for three thousand years to convince men that if an act is evil for its victim it cannot possibly be good for its perpetrator. The man who suffers injustice is not hurt as much as is the man who commits the injustice.

I cannot hope to convince men whom Plato has left indifferent. Plato's doctrine, however, is a somewhat more difficult one than is here proposed, for it can be convincingly shown that any kind of good that is achieved at the expense of others is of such a nature that it cannot be achieved by a large proportion of any

[2] Science, Philosophy and Religion, Third Symposium (New York, 1943), pp. 145 *et seq.*

group and is therefore by our definition not a social good. The doctrine of the English utilitarians was a somewhat vaguer suggestion of the same sort. There seems to have been in Bentham a slightly smug and unrealistic but quite genuinely benevolent desire to have men reach their own good ends in their own ways, and in John Stuart Mill and Robert Owen a still more humane tolerance. They were reaching for a calculus of this kind and were defeated, I think, because they could not shake loose from hidden absolutes. In what we are considering here there is also, as distinguished from Utilitarianism, a much greater insistence on the necessity, in achieving freedom, of nurturing a diversity of patterns in each cultural organization and a similar diversity among the group patterns in the whole organization of the world.

If the measure of the success of a society is the extent to which it enables men to realize the values it teaches them to reach for, then all rational and sincere service for others may be listed among the values, and any good at the expense of others is by definition doubtful. This, of course, answers only one part of the criticism. There is still the accusation that such a measure of values sanctions such values as are pursued in wars of conquest, or in any other kind of injury committed by one social group on the members of another, if they can get away with it. Half of this accusation is truth; groups of men do succeed in committing crimes against nations and much of the misery in the world is made by wickedness, not by the sorrowful accidents of the impersonal natural world. This is true and if we could believe that any social group, any tribe or nation, ever made the mere committing of crimes against their neighbors a social ideal, something which they taught their children to desire and for whose realization they created the instruments, then the other half might be true also. But that is beyond belief.

The criticism of my calculus is that some nations do succeed in conquest. True. And conquest is evil. And therefore they have succeeded in an evil purpose and thus they have proved that a nation can be successful in its purposes even when the purposes are evil. The answer is that the destructiveness and violence and disaster which it inflicts on its victims are never the purpose of an aggressive nation. They are instrumental. Even the paranoid person, or the members of a group of paranoid tendencies, use violence as a means. In politics, the means is always more important than the end. The aggressive nations commit their crimes because they want material prosperity or glory, but that also is instrumental, because they want as an ultimate and real goal some kind of happiness. The evidence will support this contention. The aggressor nations do not get happiness out of conquest for long; they generally fail in the conquest, their immediate purpose, and if they win that, the social structure built on conquest crumbles quickly. Force and violence are, like all other evils, essentially self-defeating. The good is what perpetuates itself and what perpetuates itself is good.

A free society would have the greatest possible diversity of institutions and chances for loyal attachment. It is difficult without further experience with such a society to say what the range of diversities would be. We would expect, of course, a limit in economy of time and even of material productiveness. A technological society could not work if every man had his own private system for each of his interests; this would not be a society at all. On the other hand, too much mass production of institutions and culture patterns would sacrifice our desired good to economy of energy. We would still have to maintain what I have called the democracy of culture.

There has not been much diversity in any of the societies of the world heretofore. We are deluded into thinking that peasant or primitive cultures are full of picturesque diversities. Savage life is rigidly simple to the savage, and picturesqueness is a simplicity that looks like colorfulness only to the traveler who sees the mosaic of simplicities in separate units. We have little experience to go on when we talk of the degree of pluralism that can be effectively encouraged in our free society. The doctrine stands, however, that we want as much as we can get.

Such pluralism will allow for a very wide range of normal patterns. Most men, as we have seen, will take on a simple pattern of choices without conscious deprivation or distortion. Most men can be easily educated into almost any conceivable rigidity of structure. But this necessarily costs the social group whatever amount of creative energy might have been generated by the small differences, latent in their different temperaments, that might have been brought out in a more favorable social climate.

A good society will also permit its members to change, without great cost or suffering, their own individual patterns of loyalties, shifting from one institutional group to another, not only at a time when the whole culture is making a change but even as the idiosyncrasy of personal experience. Thus the person and the culture, together or in separate ways, can find an accessible range of choices and inexpensive change because the institutions are varied and institutional loyalties are not rigidly held.

It may be urged that such a society would not be stable. This, I believe, is a fallacy. It would have the stability of a fluid movement rather than the intermittent stability of a series of explosions between moments of quiet, which well describes the social stabilities we are most familiar with. Stability, in any case, is not a good in itself although useful to our purpose, freedom.

A society that has a new concept and ideal of stability may have also a new ideal of security. There is the same difficulty in discussing security as is discovered in thinking about the democratic ideal of respect for the human personality. Security from fear and want and restraint—it is easy enough to see the evils to be fended off. But since evils are natural events judged by a subjective scale of values, we must ask not only security from what? but also, security for whom? For what kind of person and for what one of the selves of any particular person? Am I to ask security for the lazy and intemperate side of my nature, that it shall never be disturbed? Or for my inordinate ambitions? Or my secret malice? Ideals such as security, or the dignity of the personality, or similar old slogans of political progress, are too easily conceived in external terms only, and in that conception have no real meaning and no rational hope of fulfillment.

The society that we have been discussing here is one in which there will be not freedom for the sake of security as much as security for free persons, for the freedom loving qualities of its members, for the institutions and agencies that make freedom possible.

Such a society would make it simple and inexpensive in energy or social friction for any member to change his own pattern of loyalties and his way of life, but there is no reason to fear that this would mean an indefiniteness or lack of character in persons. The limit must be found here also by experience. We know that we cannot have a society in which there is security for all kinds of persons, nor for all the different changes that any person might want to make in himself. A group in which all kinds of persons were tolerated would be one in which many of them, in the natural expression of their natures, would be interfering with or preying on each other. Perhaps we can say that we want as many kinds as we can get without having any that prey on others. We want tolerance for everyone but no oppressors of any kind or size.

II

There are loud voices now crying that we have had too much science and are paying for it. To me it seems that they are saying meaningless things or else are lying by the book. It needs only the most careless look around to discover that men have been pitifully timid in using their scientific powers to direct their own lives. The major evils of our time, violence and selfishness, have indeed been the major evils of most of human history. No one can with any authority say we have got worse except that, having bigger machines, we can kill more people at a time and tell lies to greater numbers of willing fools. But the evil political creeds of today are the very denial of scientific morals in their myths and dogma. They are all regressions. No one culture group for any length of time has ever tried to create, by precise knowledge applied bravely to practical action, the conditions in which the spirit of man could realize its greatness. Societies have acted sometimes as if man could diminish himself by knowledge of himself and nature. Men have acted as if their brains were the enemies of their souls, although Pascal, in the agony of his search after salvation said, "Let us labor then to think clearly, for there is the principle of morality."

I have had much to say of freedom because it is a good that is genial to the method herein described of arriving at ideals. It is a good that can, by our calculus of a good society, be attained and perpetuated; it tends to maintain itself. But our thesis is that any other valid good can be reached for in the same way, by training men in the habits by which the good is enacted, by training them in loyalty to the institutions that make it possible, by persuading them to the changes that are later seen to be useful. Is this then a violation of the Kantian ethic, that man must not be used as means but always as an end in himself? No, unless all education and persuasion are by that measure immoral, and all spiritual encouragement to others an intrusion. The dilemma is easy to see. We either train children and persuade them to live by the ideals we believe in, or we leave them to be beggars in their own culture.

Is it anything new? Not essentially, except as a plea for seriousness in the most serious business of life. We have proved now that the world of nature, aside from man, can be understood, which is another way of saying mastered, by thinking on a series of levels of abstraction, leading at last to high reaches of symbolic play where ideal entities are moved about in the free imagination. This is not a new argument, but it is one urgently renewed, that this way of working can help also in mastering ourselves. We can, I believe, think about men as we have thought about other embodiments of energy. We can think about things abstracted out of the observation of human behavior that will be as real as atoms. We can make predictions, in time, that will be as certain and as decisive as the calculations of nuclear fission and the atomic bomb. No one can say how much this way of thinking—not a substitute for any valid tool now used but one added—will increase our powers. It may not make as much difference in the management of men as it has in the management of things, where it has made the differences between oxcarts and airplanes. It is not necessary to know what it will do. It is necessary only to try. Other methods have not done enough. And above all else it must be attempted because it is the most difficult venture that men have ever undertaken. The highest morality to a creature who can think, the inescapable duty, is to think in the most difficult regions about the most subtle and terrifying problems. Man's toughest problem is himself.

INDEX

F

G

H

I

Bei Fragen zur Produktsicherheit wenden Sie sich bitte an:
If you have any questions regarding product safety,
please contact:

Walter de Gruyter GmbH
Genthiner Straße 13
10785 Berlin
productsafety@degruyterbrill.com